Catalytic Methods in Flow Chemistry

Catalytic Methods in Flow Chemistry

Special Issue Editors
Christophe Len
Renzo Luisi

MDPI • Basel • Beijing • Wuhan • Barcelona • Belgrade

Special Issue Editors
Christophe Len
Chimie ParisTech—CNRS,
Institute of Chem. For
Life & Health Sciences
France

Renzo Luisi
Department of Pharmacy—Drug
Sciences, University of
Bari "A. Moro"
Italy

Editorial Office
MDPI
St. Alban-Anlage 66
4052 Basel, Switzerland

This is a reprint of articles from the Special Issue published online in the open access journal *Catalysts* (ISSN 2073-4344) from 2018 to 2019 (available at: https://www.mdpi.com/journal/catalysts/special_issues/catal_flow).

For citation purposes, cite each article independently as indicated on the article page online and as indicated below:

LastName, A.A.; LastName, B.B.; LastName, C.C. Article Title. *Journal Name* **Year**, *Article Number*, Page Range.

ISBN 978-3-03928-732-1 (Pbk)
ISBN 978-3-03928-733-8 (PDF)

© 2020 by the authors. Articles in this book are Open Access and distributed under the Creative Commons Attribution (CC BY) license, which allows users to download, copy and build upon published articles, as long as the author and publisher are properly credited, which ensures maximum dissemination and a wider impact of our publications.

The book as a whole is distributed by MDPI under the terms and conditions of the Creative Commons license CC BY-NC-ND.

Contents

About the Special Issue Editors . . . . **vii**

Christophe Len and Renzo Luisi
Catalytic Methods in Flow Chemistry
Reprinted from: *Catalysts* 2019, 9, 663, doi:10.3390/catal9080663 . . . . **1**

Laura Daviot, Thomas Len, Carol Sze Ki Lin and Christophe Len
Microwave-Assisted Homogeneous Acid Catalysis and Chemoenzymatic Synthesis of Dialkyl Succinate in a Flow Reactor
Reprinted from: *Catalysts* 2019, 9, 272, doi:10.3390/catal9030272 . . . . **4**

Noelia Lázaro, Ana Franco, Weiyi Ouyang, Alina M. Balu, Antonio A. Romero, Rafael Luque and Antonio Pineda
Continuous-Flow Hydrogenation of Methyl Levulinate Promoted by Zr-Based Mesoporous Materials
Reprinted from: *Catalysts* 2019, 9, 142, doi:10.3390/catal9020142 . . . . **14**

Michael Burkholder, Stanley E. Gilliland III, Adam Luxon, Christina Tang and B. Frank Gupton
Improving Productivity of Multiphase Flow Aerobic Oxidation Using a Tube-in-Tube Membrane Contactor
Reprinted from: *Catalysts* 2019, 9, 95, doi:10.3390/catal9010095 . . . . **27**

Jun Xiong and Ying Ma
Catalytic Hydrodechlorination of Chlorophenols in a Continuous Flow Pd/CNT-Ni Foam Micro Reactor Using Formic Acid as a Hydrogen Source
Reprinted from: *Catalysts* 2019, 9, 77, doi:10.3390/catal9010077 . . . . **36**

Virginie Liautard, Mélodie Birepinte, Camille Bettoli and Mathieu Pucheault
Mg-Catalyzed OPPenauer Oxidation—Application to the Flow Synthesis of a Natural Pheromone
Reprinted from: *Catalysts* 2018, 8, 529, doi:10.3390/catal8110529 . . . . **47**

Xiaojia Wang, Baosheng Jin, Hao Liu, Bo Zhang and Yong Zhang
Prediction of In-Situ Gasification Chemical Looping Combustion Effects of Operating Conditions
Reprinted from: *Catalysts* 2018, 8, 526, doi:10.3390/catal8110526 . . . . **58**

Yunshan Dong, Zongliang Qiao, Fengqi Si, Bo Zhang, Cong Yu and Xiaoming Jiang
A Novel Method for the Prediction of Erosion Evolution Process Based on Dynamic Mesh and Its Applications
Reprinted from: *Catalysts* 2018, 8, 432, doi:10.3390/catal8100432 . . . . **80**

Alexandra Gimbernat, Marie Guehl, Nicolas Lopes Ferreira, Egon Heuson, Pascal Dhulster, Mickael Capron, Franck Dumeignil, Damien Delcroix, Jean-Sébastien Girardon and Rénato Froidevaux
From a Sequential Chemo-Enzymatic Approach to a Continuous Process for HMF Production from Glucose
Reprinted from: *Catalysts* 2018, 8, 335, doi:10.3390/catal8080335 . . . . **96**

Katarzyna Maresz, Agnieszka Ciemięga and Julita Mrowiec-Białoń
Selective Reduction of Ketones and Aldehydes Microreactor—Kinetic in Continuous-Flow Studies
Reprinted from: *Catalysts* **2018**, *8*, 221, doi:10.3390/catal8050221 . . . . **116**

Claudia Carlucci, Leonardo Degennaro and Renzo Luisi
Titanium Dioxide as a Catalyst in Biodiesel Production
Reprinted from: *Catalysts* **2019**, *9*, 75, doi:10.3390/catal9010075 . . . . **124**

About the Special Issue Editors

Christophe Len received his Ph.D. from the University of Picardie-Jules Verne (UPJV) in Amiens (France) under the supervision of Professor P. Villa in the field of carbohydrate chemistry. In 1996, he joined Doctor G. Mackenzie's group at the University of Hull (UK) as a post-doctoral fellow to work on the synthesis of nucleoside analogues. In 1997, C.L. became Assistant Professor at the Laboratoire des Glucides of UPJV and worked on the chemistry of antiviral nucleoside analogues specializing in those with novel glycone systems. In 2003, C.L. received his habilitation and was promoted to full Professor in 2004 at the University of Poitiers (France), where he was the supervisor of the Biomolecules: Synthesis and Molecular Organization tem. For 2008, C.L. recieved a secondment to the University of Technology of Compiègne—UTC (France) where he created a new scientific team called Organic Chemistry and Alternative Technologies with the aim of researching green chemistry and sustainable development. In 2010, C.L. became full Professor at the University of Compiègne—UTC (France). Since November 2017, CL has developed his research at Chimie ParisTech (France) in the group Catalysis, Synthesis of Biomolecules and Sustainable Development. C.L .is a member of the French Research Network GDR2053 CNRS Continuous Flow Chemistry.Len's principal research interest is organic synthesis, including: (i) the reactivity of biomass-derived molecules (glycerol, fatty acids, lignin, cellulose) for the formation of new chemical bonds; (ii) carbon–carbon cross coupling reactions (Suzuki–Miyaura, Tsuji–Trost reactions, etc.) catalyzed by palladium complexes in aqueous media with or without ligands; (iii) homogeneous, heterogeneous, and micellar catalysis using either commercial surfactants or new bio-based ones; (iv) the use of unconventional media (water, critical fluids, ionic liquids); and (v) the use of alternative methods such as microwave irradiation, ultra-sounds, photochemistry, and continuous flow. His scientific work has been published in ca. 200 original international publications and review articles, 9 chapters, and 10 patents including publications in high profile journals. Among recent awards and recognition of his scientific career, C.L. is: Fellow of Royal Society of Chemistry (FRSC, 2015); Fellow of Association of Carbohydrate Chemists and Technologists (ACCTI, 2015); Honorary Life Fellow of Indian Society of Chemists and Biologists (ISCB, 2014); Honorary Professor of the University of Hull, England (2012–2018); and received the ACI/NBB Glycerine Innovation Research Award (2017). C.L. is member of the Editorial Board of Molecules, Catalysts, Current Green Chemistry, Sustainability, and Scientific Reports-Nature.

Renzo Luisi received a Ph.D. in Chemical Sciences at the University of Bari in 2000 under the supervision of Prof. Saverio Florio. In 2001, he was appointed research assistant at the University of Bari and in 2005, Associate Professor of Organic Chemistry at the same university. After 15 years in this role, in 2020, he was appointed full professor of Organic Chemistry at the Department of Pharmacy—Drug Sciences of the University of Bari "A. Moro". In 1999, R.L. was a Visiting Scholar to the Roger Adams Laboratory at the University of Illinois at Urbana-Champaign, working in the group of Prof. Peter Beak; in 2012, as visiting professor at the University of Manchester working in the group of Prof. Jonathan Clayden. During his career, he has been visiting scientist in several academic institutions and industries. R.L.'s main research interests revolve around organolithium-mediated stereoselective syntheses, the chemistry of heterosubstituted carbanions and mecanistic investigations using advanced spectroscopic techniques. In 2014, R.L. created the Flow Chemistry and Microreactor Technology Lab (FLAME-Lab) at the University of Bari, and started research projects in the field of sustainable chemistry using enabling technology. RL is funder of an academic spin-off, member of Italina Chemical Society and Fellow of the Royal Society of Chemistry. In 2014, R.L. was recipient of the CIMPIS Award on Innovation in Chemistry for his contribution to the field of flow chemistry based on highly reactive intermediates. In recent years, R.L. has contributed to the field of nitrene-based chemistry developing synthetic strategies for the synthesis of nitrogenated molecules.

R.L.'s research achievements are collected in more than 130 publications in international research journals, 1 book, several book chapters, and review articles. R.L. holds several international scientific collaboration with pharma companies and editorial activities. Find more at: www.renzoluisi-lab.com.

Editorial
Catalytic Methods in Flow Chemistry

Christophe Len [1],* and Renzo Luisi [2],*

1 PSL Research University, CNRS Chimie ParisTech, 11 rue Pierre et Marie Curie, F-75005 Paris, France
2 Department of Pharmacy—Drug Sciences, University of Bari "A. Moro" via E. Orabona 4, 70125 Bari, Italy
* Correspondence: christophe.len@chimieparistech.psl.eu (C.L.); renzo.luisi@uniba.it (R.L.)

Received: 14 June 2019; Accepted: 21 June 2019; Published: 2 August 2019

Continuous flow chemistry is radically changing the way of performing chemical synthesis, and several chemical and pharmaceutical companies are now investing in this enabling technology [1]. From this perspective, the development of catalytic methods in continuous flow has provided a real breakthrough in modern organic synthesis. In this Special Issue of Catalysts, recent results and novel trends are reported in the area of catalytic reactions (homogeneous, heterogeneous, and enzymatic, as well as their combinations) under continuous flow conditions. Contributions to this Special Issue include original research articles, as well as a review from experts in the field of catalysis and flow chemistry.

Two new technologies were developed, and compared, for the preparation of dialkyl succinates [2]. In particular, with the aid of homogenous acid catalysis, the trans-esterification of dimethyl succinate was achieved by using a microwave-assisted flow reactor. The use of enzymatic catalysis (with lipase Cal B) under flow conditions allowed for the preparation of dialkyl succinates by trans-esterification of dimethyl succinate. The advantages of flow reactors compared to traditional batch settings were demonstrated in this esterification process.

An innovative continuous flow process for the production of valuable 5-hydroxymethylfurfural (HMF) from glucose was developed [3]. The process proceeds via enzymatic isomerization of glucose, selective arylboronic acid-mediated fructose complexation/transportation, and a chemical dehydration to HMF. Interestingly, the new reactor was based on two aqueous phases dynamically connected via an organic liquid membrane, which enabled substantial enhancement of glucose conversion while avoiding intermediate separation steps. The use of an immobilized glucose isomerase and an acidic resin facilitated catalyst recycling.

Zirconium-based mesoporous materials were prepared and used as suitable catalysts for the continuous flow hydrogenation of methyl levulinate [4]. The catalysts were accurately characterized in order to ascertain the structure, texture, and acidic properties. All the prepared materials were successfully employed, under flow conditions, for the hydrogenation of methyl levulinate using 2-propanol as the hydrogen donor. Better performance was observed with catalysts possessing higher dispersion of ZrO_2 particles.

Monolithic flow microreactors were employed for studying the kinetics of the Meerwein–Ponndorf–Verley reduction of carbonyl compounds [5]. Zirconium-functionalized silica monoliths constituted the core of the reactor and performed well in promoting the reduction of cyclohexanone and other ketones and aldehydes using 2-butanol as the hydrogen donor. Important kinetic parameters and data on the reaction rates were the main output of this study.

In the context of the treatment of wastewater containing chlorinated organic pollutants, a continuous flow process for hydrodechlorination of chlorophenols was reported [6]. The process relies on the use of a Pd/carbon nanotube (CNT)-Ni foam microreactor system and formic acid as the hydrogen source. The catalytic system performed well in dechlorination reactions, and the catalyst could be regenerated by removing the absorbed phenol from the Pd catalyst surface.

An oxidative process conducted under continuous flow conditions, the Mg-catalyzed Oppenauer reaction was reported [7]. By using pivaldehyde or bromaldehyde as oxidants, and inexpensive magnesium tert-butoxide as the catalyst, several primary and secondary alcohols underwent oxidation reactions. A multigram continuous flow synthesis of the pheromone stemming from *Rhynchophorus ferrugineus* was realized using this oxidation method.

An approach for improving the productivity of multiphase catalytic reactions conducted in flow conditions was proposed [8]. A tube-in-tube membrane contactor (sparger) integrated in-line with the flow reactor was used in the aerobic oxidation of benzyl alcohol to benzaldehyde with a packed bed palladium catalyst. This technology was benchmarked in order to improve productivity, selectivity, and safety.

In the field of computational fluid dynamics, a predictive model has been presented for the simulation of the in situ Gasification Chemical Looping Combustion (iG-CLC) process in a circulating fluidized bed (CFB) riser fuel reactor [9]. Interestingly, CLC was demonstrated as a promising technology to implement CO_2 capture.

Another prediction method for estimating the erosion evolution is also described in this Special Issue [10]. The phenomenon of particle erosion is of great importance in industrial settings. The dynamic mesh technology was used to demonstrate the surface profile of erosion, and mathematical models were set up in order to consider gas motion, particle motion, particle-wall collision, and erosion.

A review dealt with catalytic methods for the production of biodiesel from renewable sources [11]. Titanium dioxide was targeted as the catalyst for the conversion, under batch and flow conditions, of triglycerides into fatty acid methyl esters (FAME), the main components of biodiesel.

We believe that the reported contributions in this Special Issue will inspire all those involved in the field of catalysis in flow conditions, providing useful hints for newcomers in this exciting and progressing field of science.

References

1. Bogdan, A.R.; Dombrowski, A.W. Emerging Trends in Flow Chemistry and Applications to the Pharmaceutical Industry. *J. Med. Chem.* **2019**, 456–472. [CrossRef] [PubMed]
2. Daviot, L.; Len, T.; Lin, C.S.Z.; Len, C. Microwave-Assisted Homogeneous Acid Catalysis and Chemoenzymatic Synthesis of Dialkyl Succinate in a Flow Reactor. *Catalysts* **2019**, *9*, 272. [CrossRef]
3. Gimbernat, A.; Guehl, M.; Ferreira, N.L.; Heuson, E.; Dhulster, P.; Capron, M.; Dumeignil, F.; Delcroix, D.; Girardon, J.S.; Froidevaux, R. From a Sequential Chemo-Enzymatic Approach to a Continuous Process for HMF Production from Glucose. *Catalysts* **2018**, *8*, 335. [CrossRef]
4. Lázaro, N.; Franco, A.; Ouyang, W.; Balu, A.M.; Romero, A.A.; Luque, R.; Pineda, A. Continuous-Flow Hydrogenation of Methyl Levulinate Promoted by Zr-Based Mesoporous Materials. *Catalysts* **2019**, *9*, 142. [CrossRef]
5. Maresz, K.; Ciemięga, A.; Mrowiec-Białoń, J. Selective Reduction of Ketones and Aldehydes in Continuous-Flow Microreactor—Kinetic Studies. *Catalysts* **2018**, *8*, 221. [CrossRef]
6. Xiong, J.; Ma, Y. Catalytic Hydrodechlorination of Chlorophenols in a Continuous Flow Pd/CNT-Ni Foam Micro Reactor Using Formic Acid as a Hydrogen Source. *Catalysts* **2019**, *9*, 77. [CrossRef]
7. Liautard, V.; Birepinte, M.; Bettoli, C.; Pucheault, M. Mg-Catalyzed OPPenauer Oxidation—Application to the Flow Synthesis of a Natural Pheromone. *Catalysts* **2018**, *8*, 529. [CrossRef]
8. Burkholder, M.; Gilliland, S.E., III; Luxon, A.; Tang, C.; Frank Gupton, B. Improving Productivity of Multiphase Flow Aerobic Oxidation Using a Tube-in-Tube Membrane Contactor. *Catalysts* **2019**, *9*, 95. [CrossRef]
9. Wang, X.; Jin, B.; Liu, H.; Zhang, B.; Zhang, Y. Prediction of In-Situ Gasification Chemical Looping Combustion Effects of Operating Conditions. *Catalysts* **2018**, *8*, 526. [CrossRef]

10. Dong, Y.; Qiao, Z.; Si, F.; Zhang, B.; Yu, C.; Jiang, X. A Novel Method for the Prediction of Erosion Evolution Process Based on Dynamic Mesh and Its Applications. *Catalysts* **2018**, *8*, 432. [CrossRef]
11. Carlucci, C.; Degennaro, L.; Luisi, R. Titanium Dioxide as a Catalyst in Biodiesel Production. *Catalysts* **2019**, *9*, 75. [CrossRef]

© 2019 by the authors. Licensee MDPI, Basel, Switzerland. This article is an open access article distributed under the terms and conditions of the Creative Commons Attribution (CC BY) license (http://creativecommons.org/licenses/by/4.0/).

Article

Microwave-Assisted Homogeneous Acid Catalysis and Chemoenzymatic Synthesis of Dialkyl Succinate in a Flow Reactor

Laura Daviot [1], Thomas Len [1], Carol Sze Ki Lin [2] and Christophe Len [1,3,*]

[1] Centre de Recherche de Royallieu, Université de Technologie Compiègne, Sorbonne Universités, Cedex BP20529, F-60205 Compiègne, France; laura.daviot@gmail.com (L.D.); thomaslen@orange.fr (T.L.)
[2] School of Energy and Environment, City University of Hong Kong, Tat Chee Avenue, Kowloon Tong, Hong Kong, China; carollin@cityu.edu.hk
[3] Institut de Recherche de Chimie Paris, PSL Research University, Chimie ParisTech, CNRS, UMR 8247, Cedex 05, F-75231 Paris, France
* Correspondence: christophe.len@chimieparistech.psl.eu; Tel.: +33-144-276-752

Received: 12 February 2019; Accepted: 8 March 2019; Published: 16 March 2019

Abstract: Two new continuous flow systems for the production of dialkyl succinates were developed via the esterification of succinic acid, and via the trans-esterification of dimethyl succinate. The first microwave-assisted continuous esterification of succinic acid with H_2SO_4 as a chemical homogeneous catalyst was successfully achieved via a single pass (ca 320 s) at 65–115 °C using a MiniFlow 200ss Sairem Technology. The first continuous trans-esterification of dimethyl succinate with lipase Cal B as an enzymatic catalyst was developed using a Syrris Asia Technology, with an optimal reaction condition of 14 min at 40 °C. Dialkyl succinates were produced with the two technologies, but higher productivity was observed for the microwave-assisted continuous esterification using chemical catalysts. The continuous flow trans-esterification demonstrated a number of advantages, but it resulted in lower yield of the target esters.

Keywords: continuous flow; dialkyl succinates; homogeneous catalysis; lipase Cal B; succinate

1. Introduction

With the depletion of oil-based resources, wood-based biomass and especially plant waste rich in lignocellulosic feedstocks appear to be the main alternatives for the production of platform molecules. Among them, succinic acid (SA) as a linear C-4 dicarboxylic acid is considered as one of the top 12 prospective building blocks derived from sugars by the US Department of Energy. SA is mainly produced via a chemical catalytic route starting from maleic acid and maleic anhydride. The use of furan-derived SA at laboratory-scale using chemical process, as well as via biotechnological process (i.e., by fermentation) have also been studied [1]. SA can be used as a precursor to produce different chemical intermediates [2], such as tetrahydrofuran [3], γ-butyrolactone [4], and 1,4-butanediol [5]. Particularly, SA ester products can be used in the chemical industry as a green solvent, or plastic and fuel additive, as well as in the pharmaceutical and cosmetic industries [6]. Different processes using chemical homogeneous catalysis [7–10], heterogeneous catalysis [11–21], and chemo-enzymatic reaction [22] have been reported in batch process, but few reports have described continuous flow dialkyl succinate synthesis [21,23]. Among the dialkyl succinates with value-added properties, dimethyl-, diethyl-, di-isobutyl-, and dioctyl succinates can be used as green solvents; dibutyl-, didecyl-, diamyl-, and diisoamyl succinates can be used as plastic and fuel additives; tocopherol, estriol, chloramphenicol, and hydrocortisone succinates as pharmaceutical ingredients; and dipropyl, diethoxyethyl, or diethylhexyl succinates in cosmetic application [1]. Processes for the production of dialkyl succinates in a batch reactor were developed in 2010 (Table 1). Among them, the use of sulfonic acid was the most described [8–11,14–17], followed by carboxylic acid [18] and phosphoric acid [19]. It is difficult to compare each result since the processes were conducted in different conditions by different groups. Nevertheless, the use of alcohol as both solvent and reagent was often in excess at temperature in the range of 25–160 °C for 25 h. Dialkyl succinates were produced in yields higher than 66%. Al_2O_3 was described as a heterogeneous catalyst at 25 °C for 48 h for the synthesis of dimethyl ester **2b** in 70% yield [20]. Among the recent reports, Zhang et al. described the continuous flow synthesis of diesters **2a**, **2b**, and **2d** in the presence of "man-made" heterogeneous catalyst in quantitative yield [21]. Moreover, Fabian et al. described the use of batch microwave radiation as alternative tool for the esterification of SA [14]. To the best of our knowledge, the chemoenzymatic production of diesters **2a–i** using both pure SA (**1**) and pure dimethylester **2b** as reactants has never been reported. Nevertheless, Delhomme et al. used crude fermentation broths produced from recombinant *Escherichia coli* for the synthesis of **2h** in the presence of lipase Cal B [22].

Table 1. Selected catalysts reported for the conversion of succinic acid (**1**) to dialkylsuccinates **2**.

2a (R = CH_2CH_3)
2b (R = CH_3)
2c ($CH_2CH_2CH_3$)
2d ($CH_2CH_2CH_2CH_3$)
2f ($CH_2(CH_2)_4CH_3$)
2g ($CH_2CH(CH_2CH_3)_2CH_2(CH_2)_2CH_3$)
2h ($CH_2(CH_2)_6CH_3$)
2i ($CH(CH_3)_2$)

Entry	Reactor	Catalyst	Reaction Conditions [a]	2	Yield of 2 (%)	Ref
1	batch	H_2SO_4	nd:2:110:18	2g	69	[8]
2	batch	H_2SO_4	nd:2.3:110:18	2f	78	[9]
3	batch	H_2SO_4	nd:2.3:110:18	2h	70	[9]
4	batch	OPP-SO_3H-1	10:50:70:6	2b	88	[15]
5	batch	SS-0.010	10:2:100:6.5	2a	94	[16]
6	batch	Glu-TsOH	100:80:80:4	2a	100	[17]
7	batch [b]	CH_3SO_3H@Al_2O_3	332,000:2:80:8	2b	97	[14]
8	batch [b]	CH_3SO_3H@Al_2O_3	332,000:2:80:8	2a	97	[14]
9	batch [b]	CH_3SO_3H@Al_2O_3	332,000:2:80:8	2c	97	[14]
10	batch [b]	CH_3SO_3H@Al_2O_3	332,000:2:80:8	2i	97	[14]
11	batch	$C_2(Mim)_2HSO_4$	2:3:60:3	2b	76	[10]
12	batch	$C_3(Mim)_2HSO_4$	2:3:60:3.5	2a	68	[10]
13	batch	$C_4(Mim)_2HSO_4$	2:4:60:4	2c	74	[10]
14	batch	N-Butyl-2,4-dinitro-anilinium p-toluenesulfonate	1:2:99:25	2h	93	[7]
15	batch	nano-SO_4 2-/TiO_2	5:3:160:2	2g	97	[11]
16	batch	TSA_3/MCM-41	0.1:3:80:14	2a	66	[18]
17	batch	TSA_3/MCM-41	0.1:3:80:14	2d	90	[18]
18	batch	TPA_2/MCM-41	100:3:80:8	2d	68	[19]
19	batch	TPA_2/MCM-41	100:3:80:8	2f	68	[19]
20	batch	TPA_2/MCM-41	100:3:80:8	2h	73	[19]
21	batch	Al_2O_3	50:1.6:25:48	2b	70	[20]
22	flow	PIL-A	5:1.2:85:5	2b	100	[21]
23	flow	PIL-A	5:1.2:87:4	2a	100	[21]
24	flow	PIL-A	5:1.2:100:3.5	2d	100	[21]

[a] Reaction conditions: amount of catalysts (% w/w, in some cases unit is mg):mole ratio of alcohol/succinic acid:reaction temperature (°C):reaction time (h). [b] Microwave-assisted esterification. OPP-SO_3H-1: organic knitted porous polyaromatics with pyrene; SS-0.001: silica-supported sulfate with sulfate loading 0.001 mol; TSA_3/MCM-41: 12-tungstosilicic acid (30 wt%) anchored to MCM-41; TPA_2/MCM-41: terephthalic acid (20%) anchored to MCM-41; PIL-A: acidic poly(ionic liquid).

Recently, the use of homogeneous and heterogeneous flow systems in organic chemistry have been widely studied because of their highly efficient heat transfer compared with batch methodologies, good temperature monitoring, and enhanced mass transfer [24–33]. This innovative approach also permits the time required to progress from research to pilot scale and production to be reduced. Due to our interest in the topic of green chemistry and alternative technologies, two continuous-flow systems for the production of dialkyl succinate were envisaged to develop an intensified process. Herein, we report an efficient extension of this work in order to establish a comparison between the homogeneous acid and the enzymatic continuous flow system for the production of selected dialkyl succinates.

2. Results and Discussion

Initial batch diesterification was performed using SA (**1**, 2 M) and ethanol (10 mL) in the presence of H_2SO_4 (10 mol %) at 170 °C under microwave irradiation for the production of the corresponding diester **2a** (Table 2). In the presented work, the reaction time was determined by HPLC monitoring either until no more conversion of the starting diacid **1** was observed, or within the maximum time of one hour with magnetic stirring (600 rpm). The optimization of the reaction conditions for the esterification of SA (**1**) with both acid catalysts and enzymes was first realized with a single-variable strategy, by varying one variable at a time while keeping the others constant. For the present work, error bars represent the standard deviation of five replicates. Different Bronsted acids, including H_2SO_4, H_3PO_3, *p*-touluenesulfonic acid (PTSA) and 10-camphorsulfonic acid (CSA), were tested with a concentration of 10 mol % (Table 2, entries 1–4). CSA and H_2SO_4 gave identical yields, and for economic reasons, H_2SO_4 was selected for the following study. It should be pointed out that the use of PTSA and H_3PO_4 as acid catalysts resulted in a lower yield (77% and 50%) for the same reaction time (Table 2, entries 1 and 3). The experimental results with variation of H_2SO_4 (5–20 mol %) demonstrates that the maximum yield was obtained in the presence of 20 mol % of the acid (Table 2, entry 5). Using these conditions without catalyst, compound **2a** was obtained in a low yield (9%). The acidity of the catalysts used were different (PTSA pKa −6.5; H_2SO_4 pKa −3.0, 1.9; CSA pKa 1.2 and H_3PO_4 pKa 2.1, 7.0 and 12.0). The lack of reactivity of H_3PO_4 can be related to its low acidity compared with H_2SO_4 while PTSA with a strong acidity may favor the saponification of the ester **2a**.

Table 2. Batch microwave-assisted diethyl succinate (**2a**) synthesis by varying the nature and the amount of acid at 250 W.

Entry	Acid	[Acid] (mol %)	Yield of 2a (%) [a]	Error Bar
1	PTSA	10	77	1.48
2	CSA	10	84	1.09
3	H_3PO_4	10	50	1.52
4	H_2SO_4	10	84	0.55
5	H_2SO_4	20	87	0.84
6	H_2SO_4	30	82	1.14
7	H_2SO_4	5	70	3.36

[a] The diethyl succinate yield was calculated from gas chromatography analysis with a calibration curve. CSA: 10-camphorsulfonic acid; PTSA: *p*-touluenesulfonic acid.

Based on these previous results obtained in a batch reactor, the initial reaction using the microwave continuous flow was conducted with SA (**1**, 0.15–0.27 M) in the presence of H_2SO_4 (5–20 mol %) in ethanol. The molar concentration was more diluted in the flow device compared with the batch reactor due to viscosity. Starting from SA (**1**, 0.22 M) and H_2SO_4 (20 mol %), the temperature was fixed close to the boiling point of ethanol (75 °C) with a power input of 150 W, and the residence time was fixed at 100 s for this mixture. Conversion of SA (**1**) and the yield of diethyl succinate (**2a**) were 45% and

32%, respectively. In order to improve the process, residence times were increased from 100 to 400 s. The optimal residence time was obtained at 320 s with a quantitative conversion of SA (**1**) and yield of diethyl succinate (**2a**). Using a lower amount of H_2SO_4 (5 and 10 mol %) and variation of the amount of SA (**1**, 0.15 and 0.27 M) resulted in lower yields of diethyl succinate (Table 3, entries 5–8). The use of lower temperature (30 °C and 50 °C) did not lead to improvement in conversion and yield (Table 3, entries 9 and 10).

Table 3. Continuous flow microwave-assisted diethyl succinate (**2a**) synthesis by varying the amount of SA (**1**), H_2SO_4, the residence time, and the temperature at 150 W.

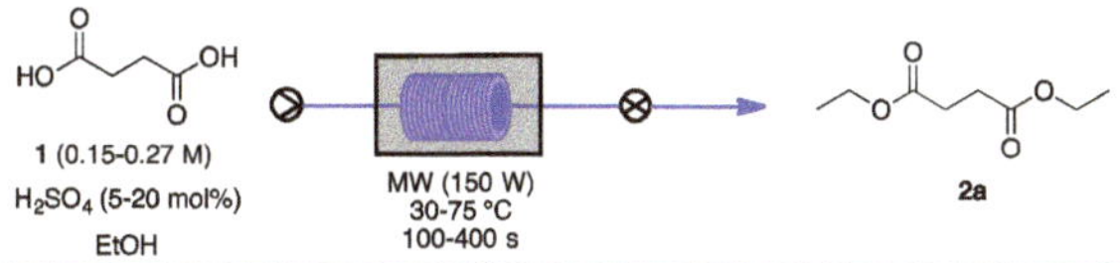

Entry	1 (mol L^{-1})	H_2SO_4 (mol %)	Temperature (°C)	Residence Time (s)	Conversion (%) [a]	Yield of 2a (%) [a]	Error Bar
1	0.22	20	75	100	45	32	0.71
2	0.22	20	75	180	60	48	1.30
3	0.22	20	75	320	100	99	0.45
4	0.22	20	75	400	100	99	0.55
5	0.22	10	75	320	85	82	0.89
6	0.22	5	75	320	75	68	0.89
7	0.27	20	75	320	95	90	2.17
8	0.15	20	75	320	82	78	1.52
9	0.22	20	50	320	73	68	1.52
10	0.22	20	30	320	35	16	2.41

[a] The diethyl succinate yield was calculated from gas chromatography analysis with a calibration curve.

Various primary and secondary alcohols having linear and branched carbon chains were subjected to the continuous esterification under our optimized conditions (Figure 1). Due to viscosity, butan-1-ol and alcohols with higher molecular weight were used at 0.18 M. Yields decreased proportionally with the increase in the number of carbons in the chain. Using primary alcohols, the conversion of SA (**1**) and yields of the selected dialkyl succinates (**2a–e**) were higher than 95% and 88%, respectively (Table 4, entries 1–5). For those primary alcohols with more than six carbon atoms, productivity decreased with yields between 65% and 80% (Table 4, entries 6–8). In contrast, the use of secondary alcohols gave similar conversion (98%) and lower yields (36% for **2i** and 89% for **2j**) (Table 4, entries 9 and 10). To the best of our knowledge, this is the first investigation which reports dialkyl succinates produced in a continuous flow. More parameters can be explored, but the present yields were similar to those obtained in the literature with batch process.

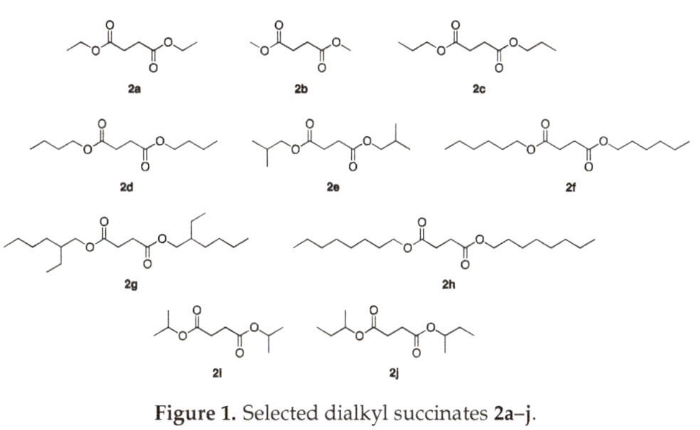

Figure 1. Selected dialkyl succinates **2a–j**.

Table 4. Scope of the microwave-assisted continuous flow dialkyl succinate **2a–j** synthesis at 75 °C.

Entry	1 (mol L^{-1})	Temperature (°C)	Conversion (%) [a]	Diesters 2	Yield of 2a–j (%) [a]	Error Bar
1	0.22	65	100	2b	100	0.89
2	0.22	75	100	2a	99	0.55
3	0.22	95	95	2c	92	0.89
4	0.18	115	98	2d	89	0.55
5	0.18	115	98	2e	88	1.30
6	0.18	115	97	2f	78	1.30
7	0.18	115	98	2g	80	4.55
8	0.18	115	96	2h	65	0.84
9	0.22	80	98	2i	89	1.64
10	0.18	96	98	2j	36	1.95

[a] The dialkyl succinate yield was calculated from gas chromatography analysis with a calibration curve.

For fair comparison, compounds **2f–h** were obtained by Stuart et al. [8,9] starting with a molar ratio of alcohol:SA (2:1) in the presence of H$_2$SO$_4$ as a catalyst at 110 °C for 18 h in a batch reactor. The yields of compounds **2f–h** were 78%, 69%, and 70%, respectively. In our optimized microwave-assisted flow synthesis, alcohols were used in large excess at similar temperature range (115 °C) for a residence time of 320 s. In this study, the yields of diesters **2f–h** were similar. It is obvious that the decrease in residence time (18 h vs. 320 s) led to significant improvement in the synthesis of biobased chemicals via esterification.

In order to explore high selectivity and smooth reaction conditions, continuous flow and bioconversion with Novozymes® 435, the lipase B from *Candida antarctica* immobilized on acrylic resin (Cal B) were studied in batch and flow reactors. The optimization of the reaction conditions for the trans-esterification of dimethylester **2b** with enzymes was realized as reported above with the acid catalysts. To probe the scope of the methodology, the influence of thermal heating, the amount of starting material **2b**, and the amount of Cal B were examined (Table 5). Dimethyl ester **2b** (50 mM) and Cal B (270 g) in ethanol were mixed in a batch reactor for 24 h by varying the temperature. Whatever the temperature used, the yield of diethyl succinate **2a** was 60% except for temperature above 60 °C due to the instability of the enzyme at high temperature (Table 5, entries 1–4). For these reasons, temperature of 20 °C was chosen and variation of the amount of enzyme was studied. For a quantity of 200 mg and 270 mg, the yields of the diesters **2a** were similar while for smaller quantities the yield of diethyl succinate **2a** were too low (Table 5, entries 5–7). The use of concentrated solution of dimethylester **2b** were tested at 20 °C in the presence of Cal B (200 mg), but the yield of diethyl succinate **2a** decreased (Table 5, entries 8 and 9).

For the transfer of the enzymatic trans-esterification from batch to continuous flow, dimethyl ester **2b** (50 mM) and Cal B (200 mg) were tested at 20 °C with different residence times (7, 2.3, and 1.2 min). The longer the time, the better the yield, regardless of the amount of the diester **2b**, enzyme dosage, and temperature (Table 6). Only dimethyl ester **2b** in the presence of a minimum amount of enzyme (200 mg) at 40 °C with a time of 7 min allowed the production of diethyl ester **2a** with a yield higher than 20% (Table 6, entries 4 and 13). It should be noted that for a doubling time of 14 min, the diester **2a** yield was 48% (Table 6, entry 23). In these optimized conditions, the use of Cal B (100 mg) resulted in only 34% (Table 6, entry 24).

Table 5. Batch chemoenzymatic synthesis of diethyl succinate (**2a**) by varying the concentration and temperature.

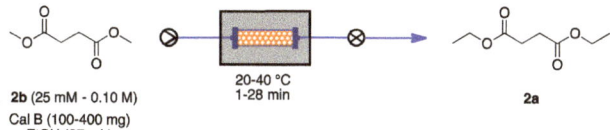

Entry	2b (M)	Cal B (mg)	Temperature (°C)	Yield of 2a (%) [a]	Error Bar
1	0.050	270	20	60	0.89
2	0.050	270	40	60	1.22
3	0.050	270	60	60	0.55
4	0.050	270	80	20	2.51
5	0.050	40	20	30	1.30
6	0.050	130	20	55	1.73
7	0.050	200	20	60	1.09
8	0.10	200	20	50	1.30
9	0.20	200	20	45	1.30

[a] The yield of diethyl succinate was calculated from gas chromatography analysis with a calibration curve.

Table 6. Flow chemoenzymatic synthesis of diethyl succinate (**2a**) by varying the concentration, temperature, and residence time.

Entry	2b (M)	Cal B (mg)	Residence Time (min)	Temperature (°C)	Conversion of 2b (%) [a]	Yield of 2a (%) [a]	Error Bar
1	0.050	200	7	20	90	7	1.00
2	0.050	200	2.3	20	79	1	0.55
3	0.050	200	1.2	20	76	1	0.45
4	0.050	200	7	40	99	23	1.30
5	0.050	200	2.3	40	73	3	0.89
6	0.050	200	1.2	40	73	2	0.84
7	0.050	200	7	60	95	18	1.22
8	0.050	200	2.3	60	78	5	1.41
9	0.500	200	1.2	60	73	2	1.00
10	0.050	100	7	40	60	14	1.09
11	0.050	100	2.3	40	68	6	0.89
12	0.050	100	1.2	40	63	traces	0.09
13	0.050	400	7	40	100	24	0.89
14	0.050	400	2.3	40	97	12	1.30
15	0.050	400	1.2	40	91	5	0.89
16	0.025	200	7	40	89	10	1.30
17	0.025	200	2.3	40	72	3	0.89
18	0.025	200	1.2	40	75	1	0.27
19	0.100	200	7	40	94	14	1.41
20	0.100	200	2.3	40	85	5	0.89
21	0.100	200	1.2	40	78	3	0.27
22	0.050	200	28	40	95	18	1.30
23	0.050	200	14	40	96	48	1.52
24	0.050	100	14	40	100	34	1.09

[a] The diethyl succinate yield was calculated from gas chromatography analysis with a calibration curve.

In order to expend the array of substrates, dimethyl ester **2b** was coupled with a variety of primary and secondary alcohols with linear and branched alkyl chains (Scheme 1). In general, the yields were twice as low as those obtained during esterification in the batch reactor. Exceptions were observed for diesters **2f**, **2g**, and **2h**, which were obtained with much lower yields. Nevertheless, the variation in yields according to the alcohol used was similar.

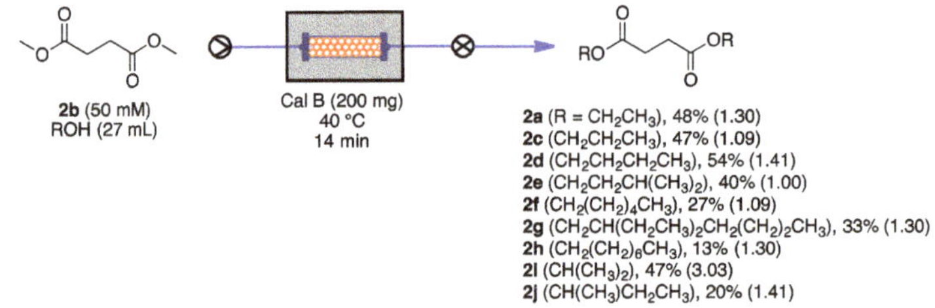

Scheme 1. Scope of the flow chemoenzymatic synthesis of dialkyl succinates **2a** and **2c–j** at 40 °C.

The selectivity of the chemoenzymatic synthesis of dialkyl succinates **2a** and **2c–j** was low using Cal B because the residence time was too low to have the second esterification. With a good conversion of the dimethylester **2b**, the first trans-esterification was obtained to furnish the intermediate and then the second trans-esterification as a limiting step gave the target compounds **2a** and **2c–j** in low-to-moderate 13%–54% yields.

3. Experimental Methods

3.1. Materials

Substrate alcohols (MeOH, EtOH, PrOH, iso-PrOH, BuOH, iso-BuOH, sec-BuOH, HexOH, 2-Et-HexOH, and OctOH) and succinic acid were purchased from Fisher Scientific (Leicestershire, United Kingdom). Diethyl succinate (**2a**) was purchased from TCI Europe (Zwijndrecht, Belgium); dimethyl succinate (**2b**), dipropyl succinate (**2c**), dibutyl succinate (**2d**), and diisopropyl succinate (**2i**) were purchased from Sigma-Aldrich (Saint Louis, MO, USA). Diisobutyl succinate (**2e**) and di-sec-butyl succinate (**2j**) were purchased from AKos Consulting & Solutions GmbH (Steinen, Germany). Dihexyl succinate (**2f**), diethylhexyl succinate (**2g**), and dioctyl succinate (**2h**) were purchased from Hangzhou DayangChem Co. Ltd. (Hangzhou, China), BOC Sciences (Shirley, NY, USA), and Carbosynth Europe (Berkshire, United Kingdom), respectively. All materials were used without purification.

3.2. Microwave-Assisted Continuous Chemical Esterification

In a typical experiment, a 500-mL Erlenmeyer flask was first filled with succinic acid (**1**, 6.50 g, 55.1 mmol, 1 equiv.) and H_2SO_4 (1.08 g, 11.1 mmol, 0.2 equiv.) in alcohol (250–300 mL). The mixture was stirred at room temperature for 10 min, and it was pumped with a peristatic pump (5 tr·min^{-1}). The solution was passed through a reactor under microwave activation (MiniFlow 200ss, Sairem®) at 65–115 °C (150 W) with a residence time of 320 s. Among the outlet solution, one milliliter of mixture was collected, pH was adjusted to 7 by washing the mixture with 5% NaOH (0.5 mL), followed by water (0.5 mL) and saturated aqueous NaCl solution (0.5 mL). Then, the organic layer was dried over anhydrous Na_2SO_4 and the solvent was removed under reduced pressure. The aqueous phase was analyzed by HPLC in order to determine the remaining succinic acid concentration, and the organic phase was analyzed by gas chromatography to quantify the amount of esters produced.

3.3. Continuous Biochemical Trans-Esterification

In a typical experiment, a solution containing dimethyl succinate (**2b**, 200 mg, 1.37 mmol, 1 equiv.) in alcohol (27 mL) was pumped at 0.05 mL min^{-1} using Syrris Asia equipment (Syrris, England). The solution was passed through a cartridge filled with Cal B (200 mg) at 40 °C, leading to a residence time of 14 min. Among the outlet solution, one milliliter of mixture was collected and saturated aqueous NaCl solution (1 mL) was added. Then, the organic layer was dried over anhydrous Na_2SO_4 and the solvent was removed under reduced pressure. The aqueous phase was analyzed by HPLC in order to determine the remaining succinic acid concentration, and the organic phase was analyzed by gas chromatography to quantify the amount of esters produced.

3.4. Gas Chromatography (GC) Analysis

Gas chromatography analyses of the organic phase were performed by a Perkin-Elmer gas chromatography instrument (Autosystem XL GC) (Perkin-Elmer, Singapore) using an Altech AT HT column with a detector at 300 °C, an injector at 340 °C, and a constant flow of nitrogen of 1 mL min^{-1}. The column was heated at 150 °C for 2 min, and the column temperature was then raised to 350 °C with a temperature gradient of 15 °C min^{-1} before being held at this temperature for 4.67 min. Succinic esters were identified and quantified by comparing GC retention time and peak area with their respective calibration standards.

3.5. High-Performance Liquid Chromatography (HPLC) Analysis

Liquid chromatography analyses of the aqueous phase were performed by a Hewlett-Packard 1090 HPLC using a reversed phase C18 column (Novapak 3.9 mm × 150 mm) held at 40 °C. Water/acetonitrile (ACN) mixture was used as the mobile phase (0.8 mL min^{-1}) in a gradient mode (0% ACN at t = 0 min to 60% ACN at t = 20 min to 90% ACN at t = 25 min to 0% ACN at t = 28 min). The species were identified by UV detection (Hitachi L400H, San Diego, CA, USA) at a wavelength of 210 nm. Succinic acid was identified and quantified by comparing GC retention time and peak area with their respective calibration standard.

4. Conclusions

In this study, two clean, mild, reproducible, and scalable continuous flow process for the production of different dialkyl succinates using H_2SO_4 as the homogeneous catalyst and Cal B as the heterogeneous catalyst were developed. A scope of different linear and branched alcohols was successfully formulated based on the optimized protocols, leading to the target chemicals. The homogeneous protocol furnished excellent yields (≥88%) when alcohols containing less than six carbon atoms were used. One exception was observed with butan-2-ol as the reactant, which is probably due to the hindered or solubility effects. For alcohols with higher molecular weight, productivities decreased with yields between 65% and 80% even if the conversion of SA (**1**) was almost quantitative. In comparison with chemical homogenous catalysis, the chemoenzymatic protocol resulted in lower yields in the order of two at best with longer residence time (14 min vs. 5 min). The lack of reactivity must be due to the lower temperature, which is related to the low thermal stability of the enzyme. To the best of our knowledge, this is the first time that dialkyl succinates have been produced in continuous flow either by chemical catalysis or enzymatic catalysis. The first method gave excellent yields and it is possible on a larger scale. The second method requires optimization through the screening of more effective enzymes.

Author Contributions: C.L. (Carol Sze Ki Lin) and C.L. (Christophe Len) conceived and designed the experiments; L.D. and T.L. performed the experiments; L.D., T.L., C.L. (Carol Sze Ki Lin), and C.L. (Christophe Len) analyzed the data; C.L. (Christophe Len) contributed reagents/materials/analysis tools; C.L. (Carol Sze Ki Lin). and C.L. (Christophe Len) wrote the paper.

Funding: This research received no external funding.

Acknowledgments: Christophe Len is thankful to Audrey Maziere and Sophie Bruniaux for technical support.

Conflicts of Interest: The authors declare no competing financial interest.

References

1. Mazière, A.; Prinsen, P.; Garcia, A.; Luque, R.; Len, C. A review of progress in (bio)catalytic routes from/to renewable succinic acid. *Biofuels Bioprod. Bioref.* **2017**, *11*, 908–931. [CrossRef]
2. Delhomme, C.; Weuster-Botz, D.; Kuhn, F.E. Succinic acid from renewable ressources as a C-4 building-block chemical—A review of the catalytic possibilities in aqueous media. *Green Chem.* **2009**, *11*, 13–26. [CrossRef]

3. Hong, U.G.; Kim, J.K.; Lee, J.; Lee, J.K.; Song, J.H.; Yi, J.; Song, I.K. Hydrogenation of succinic acid to tetrahydrofuran (THF) over rhenium catalyst supported on H_2SO_4-treated mesoporous carbon. *J. Ind. Eng. Chem.* **2014**, *20*, 3834–3840. [CrossRef]
4. Shao, Z.; Li, C.; Di, X.; Xiao, Z.; Liang, C. Aqueous-phase hydrogenation of succinic acid to g-butyrolactone and tetrahydrofuran over Pd/C, Re/C, and Pd-Re/C catalysts. *Ind. Eng. Chem. Res.* **2014**, *53*, 9638–9645. [CrossRef]
5. Kang, K.H.; Hong, U.G.; Bang, Y.; Choi, J.H.; Kim, J.K.; Lee, K.L.; Han, S.J.; Song, I.K. Hydrogenation of succinic acid to 1,4-butanediol over Re-Ru bimetallic catalysts supported on mesoporous carbon. *Appl. Catal. A* **2015**, *490*, 153–162. [CrossRef]
6. Delhomme, C. Process Integration of Fermentation and Catalysis for the Production of Succinic Acid Derivatives. Ph.D. Thesis, Technische Universitat München, Institute of Biochemical Engineering, Munich, Germany, 2011.
7. Sattenapally, N.; Wang, W.; Liu, H.; Gao, Y. i-Butyl-2,4-dinitro-anilinium p-toluenesulfonate as a highly active and selective esterification catalyst. *Tetrahedron Lett.* **2013**, *54*, 6665–6668. [CrossRef] [PubMed]
8. Stuart, A.; LeCaptain, D.J.; Lee, C.Y.; Mohanty, D.K. Poly(vinyl chloride)plasticized with mixtures of succinate di-esters—Synthesis and characterization. *Eur. Polym. J.* **2013**, *49*, 2785–2791. [CrossRef]
9. Stuart, A.; McCallum, M.M.; Fan, D.; LeCaptain, D.J.; Lee, C.Y.; Mohanty, D.K. Poly(vinyl chloride) plasticized with succinate esters: Synthesis and characterization. *Polym. Bull.* **2010**, *65*, 589–598. [CrossRef]
10. Zhao, D.; Liu, M.; Zhang, J.; Li, J.; Ren, P. Synthesis, characterization, and properties of imidazole dicationic ionic liquids and their application in esterification. *Chem. Eng. J.* **2013**, *221*, 99–104. [CrossRef]
11. Ji, X.; Chen, Y.; Shen, Z. Nano-SO_2-/TiO_2 catalyzed eco-friendly esterification of dicarboxylic acids. *Asian J. Chem.* **2014**, *26*, 5769–5772. [CrossRef]
12. Budarin, V.L.; Clark, J.H.; Luque, R.; Macquarrie, D.J. Versatile mesoporous carbonaceous materials for acid catalysis. *Chem. Commun.* **2007**, 634–636. [CrossRef] [PubMed]
13. Clark, J.H.; Budarin, V.; Dugmore, T.; Luque, R.; Macquarrie, D.J.; Strelko, V. Catalytic performance of carbonaceous materials in the esterification of succinic acid. *Catal. Commun.* **2008**, *9*, 1709–1714. [CrossRef]
14. Fabian, L.; Gomez, M.; Kuran, J.A.C.; Moltrasio, G.; Moglioni, A. Efficient microwave-assisted esterification reaction employing methanesulfonic acid supported on alumina as catalyst. *Synth. Commun.* **2014**, *44*, 2386–2392. [CrossRef]
15. Varyambath, A.; Kim, M.R.; Kim, I. Sulfonic acid-functionalized organic knitted porous polyaromatic microspheres as heterogeneous catalysts for biodiesel production. *New J. Chem.* **2018**, *42*, 12745–12753. [CrossRef]
16. Yang, Z.W.; Niu, L.Y.; Jia, X.J.; Kang, Q.X.; Ma, Z.H.; Lei, Z.Q. Preparation of silica-supported sulfate and its application as a stable and highly active solid acid catalyst. *Catal. Commun.* **2011**, *12*, 798–802. [CrossRef]
17. Zhang, B.; Ren, J.; Liu, X.; Guo, Y.; Guo, Y.; Lu, G.; Wang, Y. Novel sulfonated carbonaceous materials from p-toluenesulfonic acid/glucose as a high-performance solid-acid catalyst. *Catal. Commun.* **2010**, *11*, 629–632. [CrossRef]
18. Brahmkhatri, V.; Patel, A. Synthesis and characterization of 12-tungstosilicic acid anchored to MCM-41 as well as its use as environmentally benign catalyst for synthesis of succinate and malonate diesters. *Ind. Eng. Chem. Res.* **2011**, *50*, 13693–13702. [CrossRef]
19. Brahmkhatri, V.; Patel, A. Esterification of bioplatform molecules over 12-tungstophosphoric acid anchored to MCM-41. *J. Porous Mater.* **2013**, *20*, 209–217. [CrossRef]
20. Santacrose, V.; Bigi, F.; Casnati, A.; Maggi, R.; Storaro, L.; Moretti, E.; Vaccaro, L.; Maestri, G. Selective monomethyl esterification of linear dicarboxylic acids with bifunctional alumina catalysts. *Green Chem.* **2016**, *18*, 5764–5768. [CrossRef]
21. Zhang, J.; Zhang, S.; Han, J.; Hu, Y.; Yan, R. Uniform acid poly ionic liquid based large particle and its catalytic application in esterification reaction. *Chem. Eng. J.* **2015**, *271*, 269–275. [CrossRef]
22. Delhomme, C.; Goh, S.L.M.; Kuhn, F.E.; Weuster-Botz, D. Esterification of bio-based succinic acid in biphasic systems: Comparison of chemical and biological catalysts. *J. Mol. Catal. B Enz.* **2012**, *80*, 39–47. [CrossRef]
23. Orjuela, A.; Kolah, A.; Lira, C.T.; Miller, D.J. Mixed succinic acid/acetic acid esterification with ethanol by reactive distillation. *Ind. Eng. Chem. Res.* **2011**, *50*, 9209–9220. [CrossRef]

24. Gerardy, R.; Emmanuel, N.; Toupy, T.; Kassin, V.E.; Tshibalonza, N.N.; Schmitz, M.; Monbaliu, J.C.M. Continuous flow organic chemistry: Successes and pitfalls at the interface with current societal challenges. *Eur. J. Org. Chem.* **2018**, *20–21*, 2301–2351. [CrossRef]
25. Gerardy, R.; Morodo, R.; Estager, J.; Luis, P.; Debecker, D.P.; Monbaliu, J.C.M. Sustaining the transition from petro- to biobased chemical industry with flow chemistry. *Top. Curr. Chem.* **2019**, *377*, 1. [CrossRef]
26. Cherkasov, N.; Bao, Y.; Rebrov, E. Process intensification of alkynol semihydrogenation in a tube reactor coated with a Pd/ZnO catalyst. *Catalysts* **2017**, *7*, 358. [CrossRef]
27. Bai, Y.; Cherkasov, N.; Huband, S.; Walker, D.; Walton, R.I.; Rebrov, E. Highly selective continuous flow hydrogenation of cinnamaldehyde to cinnamyl alcohol in a Pt/SiO$_2$ coated tube reactor. *Catalysts* **2018**, *8*, 58. [CrossRef]
28. Kovalenko, G.A.; Perminova, L.V.; Beklemishev, A.B.; Parmon, V.N. Heterogeneous biocatalysts prepared by immuring enzymatic active components inside silica Xerogel and nanocarbons-in-silica composites. *Catalysts* **2018**, *8*, 177. [CrossRef]
29. Carvalho, F.; Marques, M.P.C.; Fernandes, P. Sucrose hydrolysis in a bespoke capillary wall-coated microreactor. *Catalysts* **2017**, *7*, 42. [CrossRef]
30. Sotto, N.; Cazorla, C.; Villette, C.; Billamboz, M.; Len, C. Toward the sustainable synthesis of biosourced divinylglycol from glycerol. *ACS Sustain. Chem. Eng.* **2016**, *4*, 6996–7003. [CrossRef]
31. Garcia-Olmo, A.J.; Yepez, A.; Balu, A.M.; Prinsen, P.; Garcia, A.; Mazière, A.; Len, C.; Luque, R. Activity of continuous flow synthesized Pd-based nanocatalysts in the flow hydroconversion of furfural. *Tetrahedron* **2017**, *73*, 5599–5604. [CrossRef]
32. Galy, N.; Nguyen, R.; Blach, P.; Sambou, S.; Luart, D.; Len, C. Glycerol oligomerization in continuous flow reactor. *J. Ind. Eng. Chem.* **2017**, *51*, 312–318. [CrossRef]
33. Len, C.; Bruniaux, S.; Delbecq, F.; Parmar, V.S. Palladium-catalyzed Suzuki-Miyaura cross-coupling in continuous flow. *Catalysts* **2017**, *7*, 146. [CrossRef]

© 2019 by the authors. Licensee MDPI, Basel, Switzerland. This article is an open access article distributed under the terms and conditions of the Creative Commons Attribution (CC BY) license (http://creativecommons.org/licenses/by/4.0/).

Article

Continuous-Flow Hydrogenation of Methyl Levulinate Promoted by Zr-Based Mesoporous Materials

Noelia Lázaro [1], Ana Franco [1], Weiyi Ouyang [1], Alina M. Balu [1], Antonio A. Romero [1], Rafael Luque [1,2,*] and Antonio Pineda [1,*]

[1] Departamento de Química Orgánica Universidad de Córdoba, Edificio Marie Curie (C 3), Campus de Rabanales, Ctra Nnal IV-A, Km 396, E14014 Cordoba, Spain; bt2laron@uco.es (N.L.); b12frloa@uco.es (A.F.); qo2ououw@uco.es (W.O.); qo2balua@uco.es (A.M.B.); qo1rorea@uco.es (A.A.R.)

[2] Peoples Friendship University of Russia (RUDN University), 6 Miklukho-Maklaya str., 117198 Moscow, Russia

* Correspondence: q62alsor@uco.es (R.L.); q82pipia@uco.es (A.P.); Tel.: +34-957-211-050 (R.L.); +34-957-218-623 (A.P.)

Received: 3 December 2018; Accepted: 2 January 2019; Published: 2 February 2019

Abstract: Several Zr-based materials, including ZrO_2 and Zr-SBA-15, with different silicon/zirconium molar ratios, and ZrO_2/Si-SBA-15 (where SBA-15 stands for Santa Barbara Amorphous material no. 15), have been prepared as hydrogenation catalysts. The materials were characterized using different characterization techniques including X-ray diffraction (XRD), N_2 porosimetry, scanning electron microscopy (SEM/EDX), diffuse reflectance infrared Fourier transform spectroscopy (DRIFT) of pyridine adsorption and the pulsed chromatographic method using pyridine and 2,6-dimethylpyridine as probe molecules, mainly, have been employed for the characterization of the structural, textural, and acidic properties of the synthesized materials, respectively. The catalysts have been evaluated in the hydrogenation reaction of methyl levulinate using 2-propanol as hydrogen donor solvent. The reaction conditions were investigated and stablished at 30 bar system pressure with a reaction temperature of 200 °C using around 0.1 g of catalyst and a flow rate of 0.2 mL/min flow rate of a 0.3 M methyl levulinate solution in 2-propanol. All catalysts employed in this work exhibited good catalytic activities under the investigated conditions, with conversion values in the 15–89% range and, especially, selectivity to ϒ-valerolactone in the range of 76–100% (after one hour time on stream). The highest methyl levulinate conversion and selectivity was achieved by ZrO_2/Si-SBA-15 which can be explained by the higher dispersion of ZrO_2 particles together with a highest accessibility of the Zr sites as compared with other materials such as Zr-SBA-15, also investigated in this work.

Keywords: SBA-15; zirconium; methyl levulinate; ϒ-valerolactone; flow chemistry

1. Introduction

The continuous growing demand on fuel and chemicals, which traditionally have been obtained from petroleum, together with environmental and political factors, have promoted research on alternative and renewable raw materials. In this regard, lignocellulosic biomass constitutes and important renewable carbon source due to its worldwide availability and low price [1–4]. Such lignocellulosic biomass, whose main constituents are cellulose, lignin, and hemicellulose, through different chemical transformations can be transformed into valuable chemicals with different applications such as food additives, polymers, and fuels or fuels additives. Among such chemicals γ-Valerolactone (GVL) is an important chemical easily obtained lignocellulosic biomass, specifically, from the cellulose and hemicellulose fractions. Thus, GVL is an important chemical with many applications such as sustainable green solvent as well as being precursor of many important chemicals used as intermediates for the production of fuels such as pentanoic or valeric acid, polymers, and fuel additives [3].

The formation of GVL has been reported using homogeneous as well as heterogenous catalysts. Among the homogenous catalysts, Ru-based, including Ru(acac)$_3$ [5] and RuCl$_2$(PPh$_3$) [6] compounds are predominantly used for the synthesis of GVL from different substrates such as levulinic acid [5] and glucose [7] among others. Alternatively, the use of solid catalysts entails several advantages such as easy handling and recyclability in addition to the benefits related with the environment. Among the heterogeneous catalyst used to produce GVL via hydrogenation, ruthenium catalysts, mainly Ru/C [8–10] have been the most widely among other noble metals typically used as hydrogenation catalysts including Pt [11] and Pd [12] or bimetallic combinations of these metals with others. Most of these approaches for the synthesis of GVL via hydrogenation are based on the use of molecular hydrogen whose use give rise to several disadvantages such as the low solubility of hydrogen in some solvents or the safety risks that involves the employment of high hydrogen pressures [13]. Alternatively, the use of organic molecules, including, mainly, alcohols and formic acid able to provide the reaction medium with hydrogen through the catalytic transfer hydrogenation (CTH) process. Such approach has been employed in reactions related with biomass valorization process such as the furfural hydrogenation or the conversion of levulinate esters into GVL. The previously mentioned CTH process can be favored by Lewis acid sites including those generated by isomorphic substitution of silicon by metals such as Sn, Zr and Ti in zeolites [14]. Alcohols are the solvents most wide used as hydrogen donor due to their low price, green credentials, and safety in their use. There are some factors that affect the alcohol hydrogen donor efficiency, for instance, secondary alcohols are more capable to transfer hydrogen than a primary alcohol, in addition the alcohol chain length is an important factor, thus, as longer is that chain more favored is the hydrogen transference to the substrate. Such CTH process may take place through different mechanisms that is going to depend on the catalytic species involved in the reaction: homogeneous vs. heterogeneous catalysts or which acid-base or metal catalysts are used for the reaction [13]. In this sense, Zr-based materials represent an efficient and cheap alternative to noble metals leading to high GVL yield comparable to that achieved by more expensive metals such as ruthenium and palladium among others [15,16].

Zirconium loaded materials as well as bulk zirconia are materials widely used in catalysis. Thus, our research group already reported the isomorphic substitution of silicon by zirconium in an SBA-15 framework leading to materials with highly active Lewis acid sites very active in Friedel-Crafts alkylation [17]. Similar approach has been reported for zeolites, where the replacing of aluminum by zirconium give as result a zeolite with enhanced acidic properties and resistant to aqueous environments [18]. In addition, ZrO$_2$ either supported or bulk owns interesting catalytic features, in addition to its ability of promoting CTH processes, including the combination of both acid and basic properties that allow its participation in cascade processes such as the dehydration of sugars into 5-hydroxymethylfurfural (HMF) [19] or many other applications related with biomass valorization where the water tolerance plays an important role [20,21].

In addition, continuous-flow reactors offer several advantages as compared with traditional batch reactors such the easier scaling up of the reaction conditions, possibility of testing the catalyst along large periods of time on stream in addition other advantages already reported such the high productivity, lower energy consumption, avoids the separation In this sense, continuous-flow processes may offer an attractive option for biomass valorization and the study of the catalysts under certain conditions that may occur including presence of water and moderate temperature and pressure. Moreover, our research group has demonstrated in previous studies the continuous-flow approach is successful alternative for the conversion of biomass derivatives into added value chemicals in biomass valorization processes such as the hydrogenation of furfural [22] or the production of GVL [23,24] that will be also addressed in this work. In addition to these applications, various other reactions have been reported including the continuous hydrogenation of cinammaldehyde over Pt/SiO$_2$ catalyst coated tube [25], alkynol semihydrogenation over Pd/ZnO [26] and the continuous gas phase hydrogenation of CO_2 into methanol by $CuO/ZnO/ZrO_2$ systems [27].

In this work we report the preparation of different Zr-based materials including different phases of ZrO_2, Zr-SBA-15 materials with different Si/Zr ratio prepared by direct synthesis and, finally, ZrO_2 supported on Si-SBA-15 synthesized by wet impregnation with the same loading as the Zr-SBA-15 material with the highest zirconium content, this a 10 wt % metal loading. The different aspects related with their physicochemical properties and the influence of these in the hydrogenation of methyl levulinate under continuous-flow conditions. In addition, different parameters such as flow, temperature and pressure will be optimized.

2. Results and Discussion

The textural properties of the Zr-based materials measured by nitrogen adsorption/desorption measurements are shown in Table 1. As expected, SBA-15 materials own high specific surface while ZrO_2 exhibited lower surface, more specifically the material $ZrO_2(m)$ which showed the lowest surface area (36 m^2/g). The pore diameter measured for mesoporous materials was around 6.0 nm similar to that found for the material $ZrO_2(m+c)$. By contrast, the pore size of the monoclinic zirconium oxide was much higher, so much that this material can be considered more macroporous than mesoporous. Regarding with the pore volume, as well as it happened for the specific surface, this was found higher in the SBA-15 mesoporous materials with a noticeable decrease after the incorporation of ZrO_2 due to the pore blockage by metal oxide particle in a high loading around 10 wt % and the formation ink bottle shaped pores [28].

Table 1. Textural properties of the different Zr-based materials synthesized in this work.

Material	S_{BET} (m^2/g)	V_{BJH} (cm^3/g)	V_{meso} (cm^3/g)	D_{BJH} (nm)
Zr-SBA-15(20)	651	0.79	0.72	6.1
Zr-SBA-15(10)	832	0.95	0.60	5.7
ZrO_2(m)	36	0.28	0.16	26.3
ZrO_2(m+c)	104	0.22	0.15	7.3
ZrO_2/Si-SBA-15	453	0.62	0.56	6.0

The SBA-15 materials showed isotherm plots type IV distinctive for ordered hexagonal SBA-15 materials with a hysteresis loop in a P/Po range between 0.5 and 0.8. The isotherm plot for sample loaded with Pt present a certain deterioration degree as compared with the parent material. The isotherm plots for all the synthesized materials can be found in the Supplementary Materials, Figure S1.

Low angle XRD diffractograms (Figure S2) confirmed the hexagonal arrangement characteristic for SBA-15 materials, showing peaks that can be indexed to diffraction planes 100, 110 and 200, typical for the spatial group P6mm, while this arrangement cannot be observed for the zirconium oxide materials. In addition, the wide-angle XRD of SBA-15 zirconium silicates confirmed the amorphous nature of these materials. The crystallinity of the zirconium oxides samples was evaluated using wide-angle XRD at 2θ between $10°$ and $80°$. The different phases either monoclinic or the mixture of both monoclinic and cubic present in the zirconium oxide materials can be observed on Figure 1.

While the sample $ZrO_2(m)$ shows the characteristics diffraction lines that can be indexed with zirconium oxide in the monoclinic phase, $ZrO_2(m+c)$ presents a mixture of diffraction lines corresponding with both monoclinic and cubic phases. The phases ratio was also measured for the material $ZrO_2(m+c)$, where it was found that approximately the 55% is corresponding with the cubic phase and 45% with the monoclinic phase. The particle size calculated using the Scherrer equation for $ZrO_2(m)$ was 11.6 nm quite similar to that one calculated for $ZrO_2(m+c)$ whose value was 12.1. In the case of the material $ZrO_2$10%/Si-SBA-15, despite of the high Zr loading, any diffraction line could not be detected as consequence of the high dispersion and small particle size of zirconium on the Si-SBA-15 support.

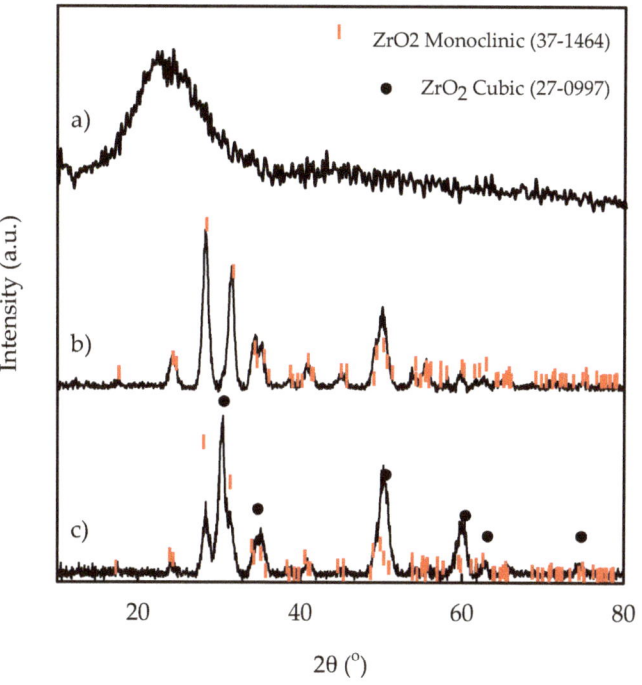

Figure 1. Wide-angle X-ray diffractograms of the synthesized ZrO_2 materials: (**a**) ZrO_2/Si-SBA 15, (**b**) ZrO_2(m) and (**c**) ZrO_2(m+c)

The surface acid properties in the zirconium loaded materials were evaluated by the chromatographic pulse method as well as by DRIFT of pyridine adsorbed. The acidity is going to be an interesting parameter to be evaluated, especially to discern the effect on the selectivity towards GVL. As it can be observed on Table 2, the incorporation of Zr in the SBA-15 framework entails and increase in total acidity, with the highest acidity value found for Zr-SBA-15(10), related with the larger zirconium loading. The material ZrO_2/Si-SBA 15 synthesized by impregnation owns higher both Lewis and Brönsted acidity, which is relate to a better accessibility of the active sites by probe molecules. This nonlinear increase in acidity is produced in Bronsted acidity while Lewis acidity remained unaltered. As expected, in the case of the bulk zirconium oxides these materials showed, mainly, Lewis acid properties together with a negligible amount of Brönsted acid sites.

Table 2. Surface acid properties of the investigated using the chromatographic pulse method.

Catalyst	Si/Zr Molar Ratio *	Total Acidity (µmol Py/g)	Brönsted Acidity (µmol DMPy/g)	Lewis Acidity (µmol/g)	B/L
Zr-SBA-15(20)	26.4	149	25	124	0.2
Zr-SBA-15(10)	12.7	223	101	122	0.8
ZrO_2(m)	-	112	<5	112	<0.1
ZrO_2(m+c)	-	149	<5	149	<0.1
$ZrO_2$10%/Si-SBA-15	12.7	271	120	151	0.8

* Determined by SEM-EDS.

The surface acid properties were measured by pyridine adsorption DRIFT for the material Zr-SBA-15(10) shows a continuous decrease in the interaction strength between pyridine moieties and Brönsted as well as Lewis acid sites with temperature (Figure 2). The characteristics bands for Brönsted and Lewis acid sites at 1550 cm^{-1} and 1442 cm^{-1}, respectively, are distinguishable even at the highest temperature measured, 300 °C. Noticeably, for the SBA-15 materials whose framework has been modified with zirconium, Brönsted acidity remains almost unaltered along the experiment, indicative of the strength of such acid sites.

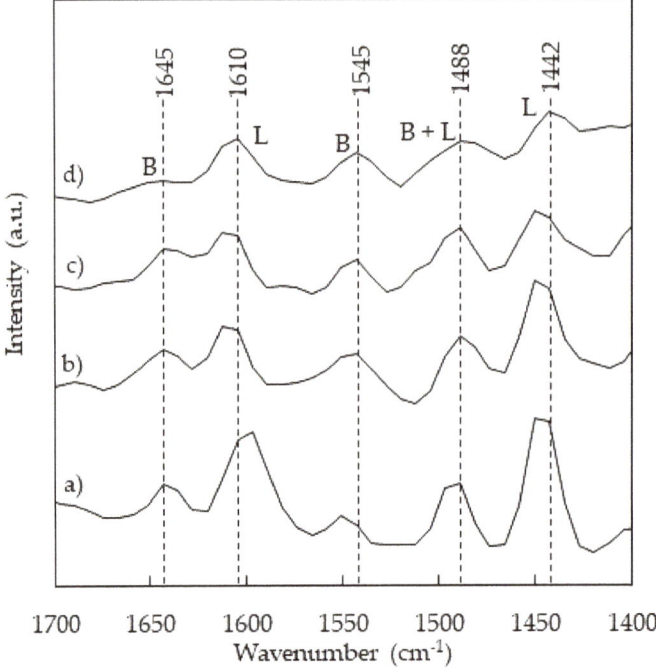

Figure 2. DRIFT spectra of pyridine adsorption on Zr-SBA-15(10) acquired at: (**a**) 100 °C, (**b**) 150 °C, (**c**) 200 °C, (**d**) 300 °C.

The catalytic activity of the Zr-based materials was evaluated in the conversion of methyl levulinate (ML) into ϒ-valerolactone using 2-propanol as H-donor solvent in a liquid-phase continuous-flow reactor (Scheme 1). In general, all the catalysts investigated showed activity in the transformation of ML into GVL. The optimum conditions were stablished for a ML concentration of 0.3 M in 2-propanol with a flow of 2 mL·min^{-1} at 200 °C temperature and the pressure was fixed at 30 bars, according to previously explored reaction conditions by our research group [23]. The first sample was withdrawn after 60 min once the steady state was achieved. The hydrogen donor solvent selected was 2-propanol that as compared with other secondary alcohols such as methanol and ethanol because of its better performance and selectivity as a previous publication of our group shows [23], that is due to the lower reduction potential of 2-propanol as compared with ethanol and methanol [29].

The results for the catalytic screening of the zirconium catalysts are shown in Figure 3. Firstly, regarding with the conversion it can be observed two different performances, while the SBA-15 materials containing Zr in their framework showed conversions below 50 mol.%, a quantitative conversion of ML for ZrO$_2$ materials in both phases and supported ZrO$_2$ on SBA-15 silicates was achieved. Remarkably, the conversion for the zirconia catalyst in the monoclinic phase deactivates continuously after the first 30 min of reaction. This deactivation may be caused by the coke deposition over the small surface of the catalyst. The results for the selectivity towards GVL for the different catalysts investigated reveals, as it happened for the conversion, a smaller selectivity for the Zr-SBA-15 materials, while the highest selectivity to GVL was achieved by both zirconium oxides studied. The decrease in selectivity for the SBA-15 zirconium silicates it is explained by the Brönsted acidity of these materials that favors the transesterification of ML to isopropyl levulinate, the secondary product detected for these catalysts as well as for the supported ZrO_2 on Si-SBA-15.

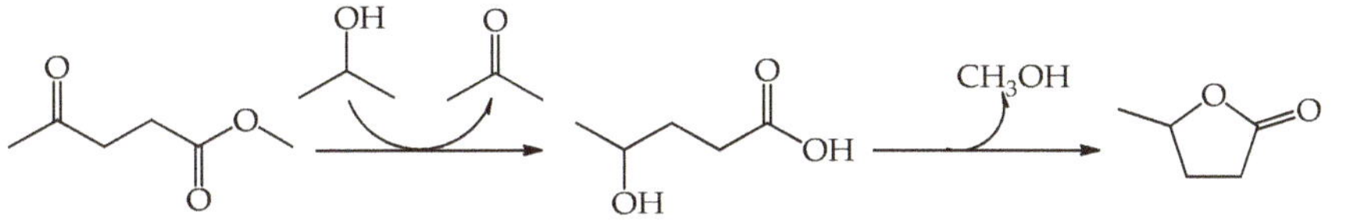

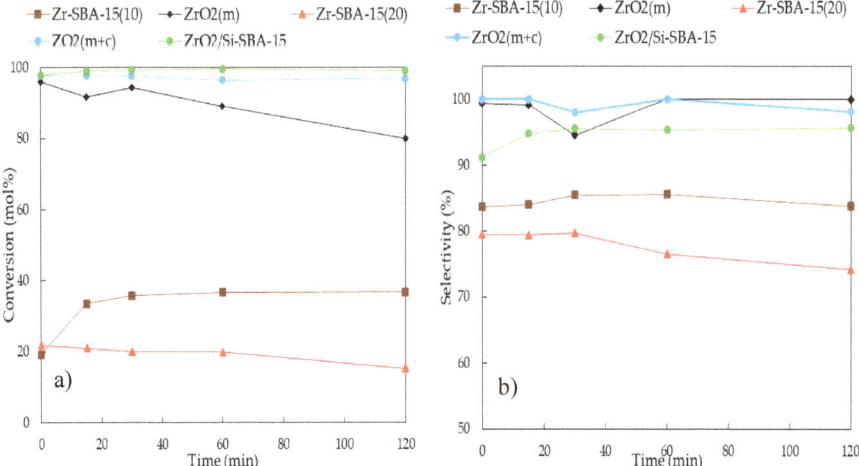

Methyl Levulinate   4-Hydroxypentanoic Acid   γ-Valerolactone(GVL)
(not observed)

Scheme 1. Methyl levulinate hydrogenation to GVL by Zr-based catalysts using 2-propanol as hydrogen donor solvent. Reproduced from Ref. [30] with permission from The Royal Society of Chemistry [30].

The differences found in terms of catalytic activity between zirconium oxide catalysts and Zr-SBA-15 materials can be attributed to the differences on Zr available to catalyze the hydrogenation step is going to be higher for ZrO_2, independently on the investigated phase. More difficult to explain are the differences found between ZrO_2/Si-SBA-15 and Zr-SBA-15(10), whose Zr loadings, theoretically, is the same. Even if it is considered that the theoretical amount of zirconium added during the preparation of the Zr-SBA-15 materials is not completely incorporated into the materials framework, the differences found in terms of conversion never could correlate with the differences in metal loading.

Figure 3. Catalytic activity of the Zr-based materials evaluated in the hydrogenation of ML to GVL expressed as: (**a**) conversion and (**b**) selectivity.

Apparently, ZrO_2 materials are the most effective materials in the conversion of ML to GVL with the highest selectivity towards the desired product. However, if the results are expressed as productivity it is possible to reach a better insight in terms of the efficiency of each material to boost the formation of GVL. Firstly, when Zr-SBA-15 materials are compared among them, the positive effect of increasing Zr content in the material framework is clear. Secondly, ZrO_2 catalyst showed similar GVL productivity to that found for Zr-SBA-15(10). Thus, the better metal dispersion expected for Zr-SBA-15 materials as compared with bulk ZrO_2 is going to play an important role in the ML hydrogenation. Finally, the most remarkable result was found when it was compared the GVL productivity (mmol of GVL produced per hour and gram of catalyst) obtained by Zr-SBA-15(10) and ZrO_2/Si-SBA-15, both with the same synthesized with the same metal loading. The GVL productivity for ZrO_2/Si-SBA-15 was almost 3 times higher the value achieved by the other materials (Figure 4).

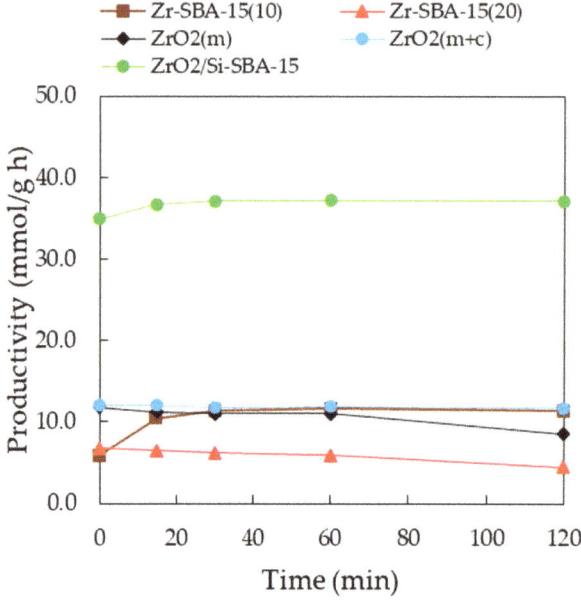

Figure 4. GVL Productivity obtained using the different zirconium materials investigated in the continuous ML hydrogenation. Reaction conditions: 0.1–0.2 g of catalyst, 0.2 mol/L ML in 2-propanol, flow rate: 2 mL/min, 30 bar, 200 °C followed during 2 h.

The differences found between Zr-SBA-15(10) and ZrO_2/Si-SBA-15 at first instance seem to be a bit contradictory. Nevertheless, Zr loading in both materials is not the only parameter that should be considered when compare the catalytic activity of both materials in the continuous ML hydrogenation. While in the material synthesized by wet impregnation (ZrO_2/Si-SBA-15) zirconium is expected to be in the external face of Si-SBA-15 channels and fully accessible by the reactant molecules, in Zr-SBA-15(10) catalyst, due to the nature of the material, an important fraction of Zr is going to be forming the walls.

Once selected the optimum catalyst for the transformation of ML to GVL over different Zr sites the influence of different experimental parameters was studied (Figure 5). Among them, pressure (10, 20 and 30 bar), temperature (120, 170, 200 bar) and flow rate (0.2, 0.5, 1 mL/min) were evaluated leaving constant the concentration of ML in 2-propanol. Firstly, the influence of temperature in the continuous hydrogenation of ML to GVL was investigated keeping constant the remaining reaction conditions (0.3 M methyl levulinate solution in 2-propanol, 0.2 mL/min flow rate, 30 bar pressure). At low temperature, 120 °C, any formation of GVL was detected while the minimum ML conversion was attributed to the transesterification of ML to produce isopropyl levulinate. Such transesterification reaction remained still important at higher temperature, 170 °C, where GVL becomes the main reaction product. A new increase in the reaction temperature impact positively in the selectivity towards GVL that is increased 95% when the reaction was performed at 200 °C as compared with the GVL selectivity obtained for the reaction at 170 °C that was 73%. Following the pressure effect was evaluated at three different pressure values (10, 20, 30 bar) controlling reaction temperature at 200 °C, 0.3 M solution of ML was introduced in the system with a speed of 0.2 mL/min. The maximum GVL productivity was achieved when a pressure of 30 bar was used as compared with the other two pressures explored. While diminishing the pressure at 20 bar leads to a slight decrease in ML conversion and GVL selectivity, a further decrease at 10 bar leads to a lower GVL productivity and increased formation of the transesterification product, isopropyl levulinate. Pressure values below 10 bar lead to low values of GVL selectivity as well as ML conversion obtaining isopropyl levulinate as main product [23]. Thus, keeping relatively high it is possible to favor the hydrogenation pathway over ML transesterification with the solvent 2-propanol. Finally, the influence of ML 0.3 M solution flow rate was evaluated at 200 °C and 30 bars. Figure 5c clearly shows that an increase on the flow rate affects ML conversion negatively, while selectivity towards GVL remained high. Thus, to achieve a good GVL productivity, for ZrO$_2$/Si-SBA-15, it is necessary to keep the flow rate low at 0.2 mL/min.

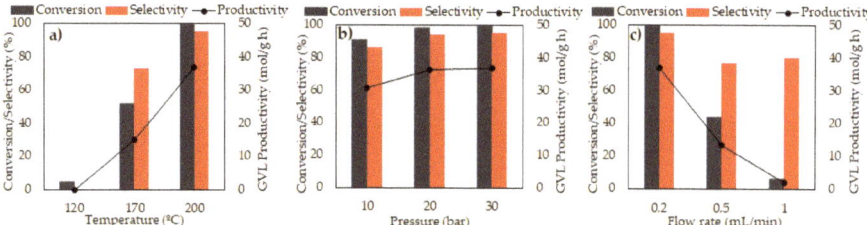

Figure 5. Catalytic performance of ZrO$_2$/Si-SBA-15 in the liquid-phase continuous transformation of ML into GVL under different reaction conditions: (**a**) temperature, (**b**) pressure, (**c**) ML solution flow rate. Reaction conditions: 0.3 M ML solution, 0.1 g of catalyst were fixed for all the experiments, (**a**) 0.2 mL/min flow rate, 30 bar pressure; (**b**) 0.2 mL/min flow rate and 200 °C reaction temperature and finally, (**c**) 30 bar pressure and 200 °C reaction temperature. Time on stream 1 h.

Additionally, a long-term test of ZrO$_2$/Si-SBA-15 for 24 h was carried out to test the catalyst stability (Figure 6). At low times on stream, below one hour, it is observed an induction period before achieving the maximum productivity. After 10 h on stream the performance of the catalyst decreased a minimal amount and then remained practically constant as proof of the stability of the catalyst under investigated conditions.

The results obtained in this research work were compared with other already published in literature, leaving room for improvement if compared with GVL productivity achieved by UiO-66 (92.3 mmol/g h) [23]. On the other hand, long-term experiments of the best material studied in this work showed a better stability with the time of ZrO$_2$/Si-SBA-15 as compared with UiO-66. Similar results were found by Rao et al. [31] who prepared several ZrO$_2$/Si-SBA-15 materials with different metal loadings using a higher amount of catalyst and harsher conditions (250 °C). Other Zr-based materials such as Zr(OH)$_4$ were evaluated in the hydrogenation of levulinate ester with poorer selectivity towards GVL (84.5%) as compared with the results herein obtained [32].

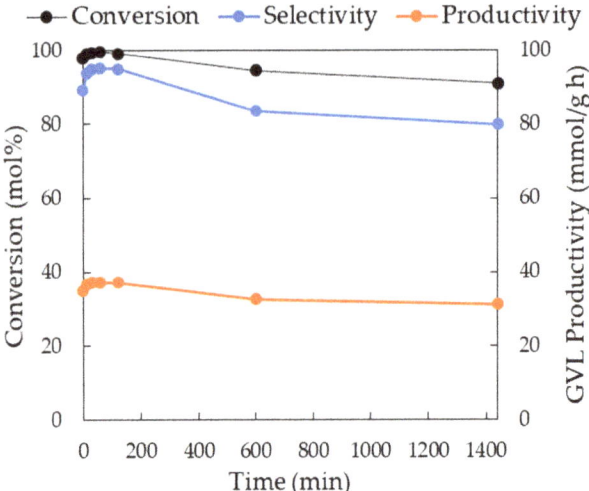

Figure 6. Long-term evaluation of ZrO$_2$/Si-SBA-15 catalytic performance in the continuous transformation of methyl levulinate into γ-valerolactone. Reaction conditions: 0.3 M ML solution, 0.1 g of catalyst, 0.2 mL/min flow rate, 30 bar, 200 °C. Time on stream up to 24 h.

3. Materials and Methods

3.1. Materials Preparation

3.1.1. Synthesis of Zr-SBA-15 Materials

Zr-modified SBA-15 materials were synthesized according to a protocol already described by our research group [17], using ZrO(NO$_3$)$_2$·xH$_2$O (Sigma-Aldrich, St. Louis, MI, USA) as metal salt precursor. For the preparation of Zr-SBA-15 materials with Si/Zr molar ratio of 20 and 10, 8 g of copolymer triblock PEO$_{20}$PPO$_{70}$PEO$_{20}$, known as Pluronic P123, (Sigma-Aldrich, St. Louis, MI, USA), used as surfactant, were dissolved in 300 mL HCl solution (pH ≈ 1.5) in a Teflon bottle. After approximately 2 h, 18 mL of tetraethyl orthosilicate (Sigma-Aldrich, St. Louis, MI, USA), used as silicon source, followed by the addition of the appropriate amount of the Zr precursor to achieve the previously mentioned molar ratios. The mixture is kept under stirring for 24 h at 35 °C and, subsequently, underwent hydrothermal treatment for 24 h. Once concluded the previous step, a white solid is formed and separated by filtration. The final zirconium silicates were obtained after calcining at 600 °C for 8 h and named as Zr-SBA-15(10) and Zr-SBA-15(20), where the numbers in brackets are the Si/Zr ratio in the synthesized materials.

3.1.2. Synthesis of Supported ZrO$_2$/Si-SBA-15

Same amount of Zr salt precursor as the used for the preparation of Zr-SBA-15(10) material was dissolved in 3 mL of distilled water and then 1 g of Si-SBA-15 was added to the zirconium containing solution. The mixture was stirred for 30 min and subsequently the water was removed using a rotary evaporator (Heidolph Laborota 4000, Schwabach, Germany). The dried solid obtained was calcined at 400 °C during 4 h with a heating rate of 2 °C/min.

3.1.3. Synthesis of ZrO$_2$

Two ZrO$_2$ oxides were synthesized at different pH values. Both were prepared following the same protocol described in Section 3.1. without adding TEOS. The samples were calcined at 600 °C for 8 h and named ZrO$_2$(m) and ZrO$_2$(m+c), where m stands for monoclinic and m+c for monoclinic and cubic.

3.2. Materials Characterization

The nitrogen adsorption/desorption isotherm have been obtained at liquid nitrogen temperature (77K) using a Micromeritics ASAP 2000 porosimeter (Micromeritics Instrument Corp., Norcross, GA, USA). The amount of sample employed for each analysis was in a range 0.18–0.23 g, and prior to the analysis all the solids were outgassed at 130 °C for 24 h. The specific surface area of the synthesized materials has been calculated using the linear part (0.05 < Po < 0.22) of Brunauer, Emmett and Teller (BET) equation. Pore size distribution have been calculated using the adsorption branch and the Barrett, Joyner, and Halenda (BJH) equation (Barret-Joyner-Halenda). Pore volume have been calculated using the BJH formula.

The crystallinity and the structure of the synthesized catalysts have been evaluated using the X-ray diffraction technique. X-ray diffractograms were acquired using a Bruker D8D Discover (40 kV, 40 mA) diffractometer (Bruker AXS, Karlsruhe, Germany) using the Cu Kα radiation (λ = 1.54 Å). The scan speed was 0.5 or 1 °/min in the interval 0.5° < 2θ < 5° for the low angle measurements and 10° < 2θ < 80° for the wide-angle acquisitions. This is instrument is fitted with the "Diffract.Suite EVA" software version 3.1 (Bruker AXS, Karlsruhe, Germany) which allows particle size measurement and phase ratio determination.

Elemental analysis of the synthesized materials was carried out using a JEOL JSM 7800F scanning electron microscope (JEOL Ltd., Akishima, Tokio, Japón) fitted with a X-max150 microanalysis system, window type detector SiLi, detection range: from boron to uranium, 127 eV resolution at 5.9 KeV.

The surface acid properties were evaluated using a chromatographic pulse method using pyridine, which measures total acidity, and 2,6-dimethyl pyridine that interact, mainly, with the Brönsted acid sites. Lewis acidity was obtained by difference between total and Brönsted acidity. The measurements were carried out a gas chromatograph fit with a flame ionization detector (FID) detector and a Chromosorb AW-MCS 80/100 packed column of 0.5 m length containing 5% wt. % in polyphenylether (Supelco Analytical, Bellefonte, PA, USA). The operational conditions for the analysis were set up as follows: inlet temperature 300 °C, FID detector at 250 °C, and oven temperature was stablished at 70 °C and 90 °C for pyridine and 2,6-dimethyl pyridine, respectively.

Additionally, the qualitative evaluation of the acidity has been analyzed using DRIFT of adsorbed pyridine in the region between 1700 and 1400 cm^{-1}. The spectra were recorded at 100, 150, 200 and 300 °C to assess the acidity strength. The vibration modes at 1545 cm^{-1} corresponds with the interaction of the pyridinium cation with Brönsted acid sites, while at ca. 1442 cm^{-1} rise the band corresponding with Lewis acid sites. An ABB 3000 instrument provided with a PIKE Technologies DiffusIR (PIKE Technologies, Madison, WI, USA), a diffuse reflection accessory that can operate at different temperatures and gas environments.

3.3. Catalytic Activity

The catalytic activity of the synthesized materials was investigated in the ML hydrogenation using 2-propanol as hydrogen donor solvent. The amount of catalyst used was approximately 0.1 g, (0.1 g for SBA-15 materials and 0.2 g for ZrO$_2$ catalysts, based on material density). The optimum conditions to perform this reaction were stablished at 200 °C, at 30 bar pressure, with a 0.2 mL/min flow of a 0.3 M ML solution in 2-propanol. Additionally, different pressure, temperature, and ML solution flow were evaluated to determine the best reaction conditions to perform this reaction under continuous flow. Alicquots were collected one hour after the steady state conditions were reached in the reactor to ensure that the entire line was filled with ML solution.

The samples were analyzed in a series II Agilent 5890 GC, provided with a SUPELCO EQUITY TM-1 (60 m × 0.25 mm × 0.25 μm) column and an FID detector (Supelco Analytical, Bellefonte, PA, USA). The temperature in the injector as well as in the reactor were 250 °C. The oven temperature program used was from an initial temperature of 60 °C for one minute that was increases up to 230 °C with a heating rate of 10 °C/min and remained constant at that temperature for 5 min. Reaction products were confirmed using a 7820A GC coupled with a 5977B mass spectrometer detector (Agilent Technologies, Santa Clara, CA, USA) using same analysis conditions as above.

4. Conclusions

Several materials based on zirconium: two SBA-15 zirconium silicates with two different Si/Zr molar ratio, two bulky ZrO_2 with different phases and ZrO_2 loaded on Si-SBA-15 (ZrO_2/Si-SBA-15) were investigated in the continuous ML hydrogenation to GVL using 2-propanol as hydrogen donor solvent. Among them, ZrO_2/Si-SBA-15 showed the best performance in terms of GVL productivity, which can be explained by, firstly by the better dispersion of the ZrO_2 particles as compared with the bulky oxides in addition to an expected lower particle size and, secondly, the better accessibility of the Zr sites in this material compared with the material Zr-SBA-15(10), with same Zr loading, where the zirconium sites is going to be in the material framework forming the SBA-15 walls. In addition, ZrO_2/Si-SBA-15 have displayed a significant stability with time on stream keeping almost constant GVL productivity with a slight decrease in selectivity and ML conversion. Also, it is noteworthy the high selectivity towards GVL achieved by all the investigated materials, with a minimal amount found of isopropyl levulinate, the transesterification product. Finally, the influence of several experimental variable on the continuous hydrogenation of ML by ZrO_2/Si-SBA-15 such as reaction temperature, system pressure and methyl levulinate solution feeding have been investigated. It was observed that high pressure and temperature affects GVL productivity positively while flow rates higher to 0.2 mL/min were too fast to convert ML efficiently.

Supplementary Materials: The following are available online at http://www.mdpi.com/2073-4344/9/2/142/s1, Figure S1: Nitrogen adsorption-desorption plots for the materials: (**a**) Zr-SBA-15(10), (**b**) ZrO2/Si-SBA-15, (**c**) Zr-SBA-15(20), (**d**) ZrO2(m). Figure S2: Low-angle XRD for selected materials: ZrO_2/Si-SBA-15, Zr-SBA-15(10), ZrO_2(m). Characteristics diffraction lines (100), (110), (200) corresponding with hexagonal arrangement are shown for SBA-15 materials.

Author Contributions: Conceptualization, R.L. and A.A.R.; methodology, A.P., A.A.R. and R.L.; investigation, N.L., A.F, W.O.; resources, A.P., A.A.R.; data curation, N.L., A.F, W.O.; writing—original draft preparation, A.P.; writing—review and editing, R.L., A.M.B.; visualization, A.P.; supervision, R.L., A.M.B., A.A.R; project administration, A.A.R, A.M.B.; funding acquisition, R.L.

Funding: This research was funded by RUDN University Program 5-100.

Acknowledgments: Antonio Pineda acknowledges the support of "Plan Propio de Investigación" from Universidad de Córdoba (Spain) and "Programa Operativo" FEDER funds from Junta de Andalucía. R.L. gratefully acknowledges funding from MINECO under project CTQ2016-78289-P, co-financed with FEDER funds, also for an FPI contract for A.F. (BES-2017-081560) under the framework of the granted project. The publication has been prepared with support from RUDN University Program 5-100.

Conflicts of Interest: The authors declare no conflict of interest.

References

1. Chheda, J.N.; Huber, G.W.; Dumesic, J.A. Liquid-Phase Catalytic Processing of Biomass-Derived Oxygenated Hydrocarbons to Fuels and Chemicals. *Angew. Chem. Int. Ed.* **2007**, *46*, 7164–7183. [CrossRef] [PubMed]
2. Climent, M.J.; Corma, A.; Iborra, S. Conversion of biomass platform molecules into fuel additives and liquid hydrocarbon fuels. *Green Chem.* **2014**, *16*, 516–547. [CrossRef]
3. Corma, A.; Iborra, S.; Velty, A. Chemical Routes for the Transformation of Biomass into Chemicals. *Chem. Rev.* **2007**, *107*, 2411–2502. [CrossRef] [PubMed]
4. Serrano-Ruiz, J.C.; Luque, R.; Sepulveda-Escribano, A. Transformations of biomass-derived platform molecules: From high added-value chemicals to fuels via aqueous-phase processing. *Chem. Soc. Rev.* **2011**, *40*, 5266–5281. [CrossRef] [PubMed]

5. Mehdi, H.; Fábos, V.; Tuba, R.; Bodor, A.; Mika, L.T.; Horváth, I.T. Integration of Homogeneous and Heterogeneous Catalytic Processes for a Multi-step Conversion of Biomass: From Sucrose to Levulinic Acid, γ-Valerolactone, 1,4-Pentanediol, 2-Methyl-tetrahydrofuran, and Alkanes. *Top. Catal.* **2008**, *48*, 49–54. [CrossRef]
6. Osakada, K.; Ikariya, T.; Yoshikawa, S. Preparation and properties of hydride triphenyl-phosphine ruthenium complexes with 3-formyl (or acyl) propionate [RuH(OCOCHRCHRCOR′)(PPh3)3] (R=H, CH3, C2H5; R=H, CH3, C6H5) and with 2-formyl (or acyl) benzoate [RuH(o-OCCOC6H4COR′)(PPh3)3] (R′=H, CH3). *J. Organomet. Chem.* **1982**, *231*, 79–90. [CrossRef]
7. Braca, G.; Maria, A.; Galletti, R.; Sbrana, G. Anionic ruthenium iodocarbonyl complexes as selective dehydroxylation catalysts in aqueous solution. *J. Organomet. Chem.* **1991**, *417*, 41–49. [CrossRef]
8. Hengne, A.M.; Biradar, N.S.; Rode, C.V. Surface Species of Supported Ruthenium Catalysts in Selective Hydrogenation of Levulinic Esters for Bio-Refinery Application. *Catal. Lett.* **2012**, *142*, 779–787. [CrossRef]
9. Tan, J.J.; Cui, J.L.; Deng, T.S.; Cui, X.J.; Ding, G.Q.; Zhu, Y.L.; Li, Y.W. Water-Promoted Hydrogenation of Levulinic Acid to γ-Valerolactone on Supported Ruthenium Catalyst. *ChemCatChem* **2015**, *7*, 508–512. [CrossRef]
10. Kuwahara, Y.; Kaburagi, W.; Fujitani, T. Catalytic transfer hydrogenation of levulinate esters to γ-valerolactone over supported ruthenium hydroxide catalysts. *RSC Adv.* **2015**, *45848*–*45855*. [CrossRef]
11. Vu, H.; Harth, F.M.; Wilde, N. Silylated Zeolites with Enhanced Hydrothermal Stability for the Aqueous-Phase Hydrogenation of Levulinic Acid to γ-Valerolactone. *Front. Chem.* **2018**, *143*. [CrossRef] [PubMed]
12. Wang, A.Q.; Lu, Y.R.; Yi, Z.X.; Ejaz, A.; Hu, K.; Zhang, L.; Yan, K. Selective Production of γ-Valerolactone and Valeric Acid in One-Pot Bifunctional Metal Catalysts. *ChemistrySelect* **2018**, *3*, 1097–1101. [CrossRef]
13. Gilkey, J.G.; Xu, B. Heterogeneous Catalytic Transfer Hydrogenation as an Effective Pathway in Biomass Upgrading. *ACS Catal.* **2016**, *6*, 1420–1436. [CrossRef]
14. Luo, H.Y.; Consoli, D.F.; Gunther, W.R.; Román-Leshkov, Y. Investigation of the reaction kinetics of isolated Lewis acid sites in Beta zeolites for the Meerwein–Ponndorf–Verley reduction of methyl levulinate to γ-valerolactone. *J. Catal.* **2014**, *320*, 198–207. [CrossRef]
15. Tang, X.; Hu, L.; Sun, Y.; Zhao, G.; Hao, W.W.; Lin, L. Conversion of biomass-derived ethyl levulinate into γ-valerolactone via hydrogen transfer from supercritical ethanol over a ZrO2 catalyst. *RSC Adv.* **2013**, *3*, 10277–10284. [CrossRef]
16. Chia, M.; Dumesic, J.A. Liquid-phase catalytic transfer hydrogenation and cyclization of levulinic acid and its esters to γ-valerolactone over metal oxide catalysts. *Chem. Commun.* **2011**, *47*, 12233–12235. [CrossRef] [PubMed]
17. Gracia, M.D.; Balu, A.M.; Campelo, J.M.; Luque, R.; Marinas, J.M.; Romero, A.A. Evidences of the in situ generation of highly active Lewis acid species on Zr-SBA-15. *Appl. Catal. A* **2009**, *371*, 85–91. [CrossRef]
18. Wolf, P.; Hammond, C.; Conrad, S.; Hermans, I. Post-synthetic preparation of Sn-, Ti- and Zr-beta: A facile route to water tolerant, highly active Lewis acidic zeolites. *Dalton Trans.* **2014**, *43*, 4514–4519. [CrossRef]
19. Chareonlimkun, A.; Champreda, V.; Shotipruk, A.; Laosiripojana, N. Catalytic conversion of sugarcane bagasse, rice husk and corncob in the presence of TiO_2, ZrO_2 and mixed-oxide TiO_2-ZrO_2 under hot compressed water (HCW) condition. *Bioresour. Technol.* **2010**, *101*, 4179–4186. [CrossRef] [PubMed]
20. Pichler, C.M.; Al-Shaal, M.G.; Gu, D.; Joshi, H.; Ciptonugroho, W.; Schüth, F. Ruthenium Supported on High-Surface-Area Zirconia as an Efficient Catalyst for the Base-Free Oxidation of 5-Hydroxymethylfurfural to 2,5-Furandicarboxylic Acid. *ChemSusChem* **2018**, *11*, 2083–2090. [CrossRef]
21. Wattanapaphawonga, P.; Reubroycharoen, P.; Yamaguchi, A. Conversion of cellulose into lactic acid using zirconium oxide catalysts. *RSC Adv.* **2017**, *7*, 18561–18568. [CrossRef]
22. Garcia-Olmo, A.J.; Yepez, A.; Balu, A.M.; Prinsen, P.; Garcia, A.; Maziere, A.; Len, C.; Luque, R. Activity of continuous flow synthesized Pd-based nanocatalysts in the flow hydroconversion of furfural. *Tetrahedron* **2017**, *73*, 5599–5604. [CrossRef]
23. Ouyang, W.; Zhao, D.; Wang, Y.; Balu, A.M.; Len, C.; Luque, R. Continuous Flow Conversion of Biomass-Derived Methyl Levulinate into γ-Valerolactone Using Functional Metal Organic Frameworks. *ACS Sustain. Chem. Eng.* **2018**, *6*, 6746–6752. [CrossRef]
24. Fu, J.; Sheng, D.; Lu, X. Hydrogenation of Levulinic Acid over Nickel Catalysts Supported on Aluminum Oxide to Prepare γ-Valerolactone. *Catalysts* **2016**, *6*, 6. [CrossRef]

25. Bai, Y.; Cherkasov, N.; Huband, S.; Walker, D.; Walton, R.; Rebrov, E. Highly Selective Continuous Flow Hydrogenation of Cinnamaldehyde to Cinnamyl Alcohol in a Pt/SiO$_2$ Coated Tube Reactor. *Catalysts* **2018**, *8*, 58. [CrossRef]
26. Cherkasov, N.; Bai, Y.; Rebrov, E. Process Intensification of Alkynol Semihydrogenation in a Tube Reactor Coated with a Pd/ZnO Catalyst. *Catalysts* **2017**, *7*, 358. [CrossRef]
27. Huang, C.; Chen, S.; Fei, X.; Liu, D.; Zhang, Y. Catalytic Hydrogenation of CO$_2$ to Methanol: Study of Synergistic Effect on Adsorption Properties of CO$_2$ and H$_2$ in CuO/ZnO/ZrO$_2$ System. *Catalysts* **2015**, *5*, 1846–1861. [CrossRef]
28. Schüth, F.; Wingen, A.; Sauer, J. Oxide Loaded Ordered Mesoporous Oxides for Catalytic Applications. *Micropor. Mesopor. Mater.* **2001**, *44–45*, 465–476. [CrossRef]
29. Li, H.; Fang, Z.; Yang, S. Direct Conversion of Sugars and Ethyl Levulinate into γ-Valerolactone with Superparamagnetic Acid–Base Bifunctional ZrFeOx Nanocatalysts. *ACS Sustain. Chem. Eng.* **2016**, *4*, 236–246. [CrossRef]
30. Hernández, B.; Iglesias, J.; Morales, G.; Paniagua, M.; López-Aguado, C.; García Fierro, J.L.; Wolf, P.; Hermans, I.; Melero, J.A. One-pot cascade transformation of xylose into γ-valerolactone (GVL) over bifunctional Brønsted–Lewis Zr–Al-beta zeolite. *Green Chem.* **2016**, *18*, 5777–5781. [CrossRef]
31. Enumula, S.S.; Gurram, V.R.B.; Kondeboina, M.; Burri, D.R.; Kamaraju, S.R.R. ZrO$_2$/SBA-15 as an efficient catalyst for the production of γ-valerolactone from biomass-derived levulinic acid in the vapour phase at atmospheric pressure. *RSC Adv.* **2016**, *6*, 20230–20239. [CrossRef]
32. Tang, X.; Chen, H.; Hu, L.; Hao, W.; Sun, Y.; Zeng, X.; Lin, L.; Liu, S. Conversion of biomass to γ-valerolactone by catalytic transfer hydrogenation of ethyl levulinate over metal hydroxides. *Appl. Catal. B* **2014**, *147*, 827–834. [CrossRef]

© 2019 by the authors. Licensee MDPI, Basel, Switzerland. This article is an open access article distributed under the terms and conditions of the Creative Commons Attribution (CC BY) license (http://creativecommons.org/licenses/by/4.0/).

Article

Improving Productivity of Multiphase Flow Aerobic Oxidation Using a Tube-in-Tube Membrane Contactor

Michael Burkholder [1], Stanley E. Gilliland III [1,2], Adam Luxon [1], Christina Tang [1,*] and B. Frank Gupton [1,*]

1. Department of Chemical and Life Science Engineering, Virginia Commonwealth University, 601 W. Main St., Richmond, VA 23284-3028, USA; mbburkholder@vcu.edu (M.B.); gillilandse@vcu.edu (S.E.G.III); aluxon@vcu.edu (A.L.)
2. Venebio Group, LLC, 7400 Beauford Springs Drive, Suite 300 Richmond, VA 23225, USA
* Correspondence: ctang2@vcu.edu (C.T.); bfgupton@vcu.edu (B.F.G.); Tel.: +1-804-827-1917 (C.T.); +1-804-828-4799 (B.F.G.)

Received: 28 November 2018; Accepted: 12 January 2019; Published: 17 January 2019

Abstract: The application of flow reactors in multiphase catalytic reactions represents a promising approach for enhancing the efficiency of this important class of chemical reactions. We developed a simple approach to improve the reactor productivity of multiphase catalytic reactions performed using a flow chemistry unit with a packed bed reactor. Specifically, a tube-in-tube membrane contactor (sparger) integrated in-line with the flow reactor has been successfully applied to the aerobic oxidation of benzyl alcohol to benzaldehyde utilizing a heterogeneous palladium catalyst in the packed bed. We examined the effect of sparger hydrodynamics on reactor productivity quantified by space time yield (STY). Implementation of the sparger, versus segmented flow achieved with the built in gas dosing module (1) increased reactor productivity 4-fold quantified by space time yield while maintaining high selectivity and (2) improved process safety as demonstrated by lower effective operating pressures.

Keywords: flow chemistry; continuous reactor; tube-in-tube; multiphase catalysis; oxidation; aerobic

1. Introduction

Multiphase catalytic reactions (e.g., hydrogenation and oxidation) are important in the production of petroleum-derived products, commodity and fine chemicals, and pharmaceuticals [1–3]. Such reactions using conventional reactors (e.g., trickle-beds, bubble columns) are challenging due to insufficient heat- and mass transfer, low interfacial areas, and potential process safety concerns [3–6]. Flow reactor technology is a promising alternative to achieve safer and more efficient on-demand manufacturing that can be scaled with parallel reactors [3–5,7–10]. Key benefits include (i) enhanced heat and mass transfer due to inherently large surface area-to-volume ratios; and (ii) small volumes which mitigate safety issues [3–5,7–11].

Performing aerobic, selective oxidation of alcohols to aldehydes using flow reactors could greatly reduce economic and environmental costs associated with chemical processing [2,8,12]. Several factors must be considered to acquire these benefits, including the type of oxidant and catalyst. Using catalytic amounts of oxygen gas as the terminal oxidant, water is typically the only byproduct and it avoids the use of stoichiometric amounts of harsh oxidizing agents [13]. Both homogeneous and heterogeneous catalysts have the ability to facilitate selective oxidations using gas oxidants; however, heterogeneous catalysts are advantageous due to the ease of product separation in continuous reactors [2,12,14–20]. In these solid-liquid-gas reactions, hydrodynamics and wetting are important considerations [5,21].

Hii and co-workers demonstrated selective oxidation of primary and secondary alcohols using a commercially available catalyst and flow reactor [19]. Using 0.1–1 M substrate concentrations of benzyl alcohol, complete conversion and >99% selectivity was achieved [19]. However, the dilute conditions inherently limited reactor productivity. Alternatively, Gavriilidis and co-workers built a tube-in-tube membrane reactor for selective oxidation of neat benzyl alcohol to benzaldehyde [22]. The role of the tube-in-tube membrane reactor was to facilitate the gas oxidant diffusion into the reacting liquid to feed to the catalytic packed bed [8,10,22–29]. They found that the tube-in-tube reactor improved the catalyst contact with both benzyl alcohol and oxygen simultaneously, resulting in increased conversion and selectivity. This result indicated that the reactor performance was limited by oxygen permeation. Diluting the catalyst bed with silica beads to increase the catalyst contact time also improved selectivity; however, the maximum selectivity obtained was less than 80% in all cases [22]. From these studies, the authors indicated that improving reactor productivity while maintaining selective oxidation was desirable.

We investigated the use of a tube-in-tube membrane contactor with a commercial continuous reactor to improve reactor productivity of a multiphase (solid-liquid-gas) catalytic reactions. Specifically, we introduced a tube-in-tube membrane contactor (sparger) to a flow reactor for examination of the aerobic oxidation of benzyl alcohol by a heterogeneous palladium catalyst using oxygen or air as a model reaction. It is important to note that the reactor is the specific focus of this investigation, rather than the choice of substrate or catalyst. The effect of the sparger on hydrodynamics and resulting selectivity and reactor productivity are discussed.

2. Results and Discussion

Selective aerobic oxidation of benzyl alcohol to benzaldehyde via a commercial palladium-based catalyst was selected as a model system to examine the effect of gas-liquid contact in heterogeneous catalytic reactions performed in flow. Figure 1 depicts the full reaction pathway for the oxidation of benzyl alcohol. Use of noble metals such as palladium (Pd) were known to facilitate aerobic oxidation of alcohols, and Pd has been shown to be more active than Pt or Ru for selective oxidation of benzyl alcohol [30]. Furthermore, the mechanism of the reaction on the Pd surface has been well characterized [16]. Even with an active Pd catalyst, achieving selective aldehyde production was challenging since the thermodynamics favor the over-oxidized products [16,21]. The catalyst and reactor conditions affected the conversion and selectivity to the desired aldehyde product [21].

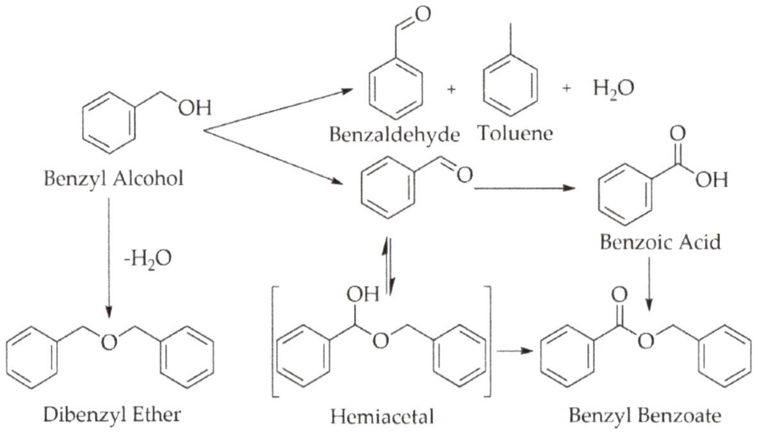

Figure 1. Reaction mechanism for the oxidation of benzyl alcohol.

The reaction was first carried out continuously using the ThalesNano X-Cube™ reactor (Figure S1, Budapest, Hungary). An external reciprocating syringe pump upstream of an HPLC pump was added for the liquid stream to achieve precise residence times in the packed bed. The gas dosing module within the X-Cube™ maintained segmented flow with a fixed 1:19 volumetric gas bubble to liquid slug ratio throughout the process. In these initial experiments, the oxidation of benzyl alcohol was examined over 8 min of total residence time with 10 passes while monitoring the conversion of benzyl alcohol and selectivity to benzaldehyde. With air as the oxidant, the highest conversion of benzyl alcohol was 2.1% achieved after 8 passes with selectivity averaging 76% across the process (Figure S2).

To determine if availability of oxygen at the catalyst surface was limiting conversion, oxygen was used as the oxidant in place of air. The conversion of benzyl alcohol using oxygen was 60% higher than air: 3.4% compared to 2.1% after 6 min of residence time (Figure S2). Similarly, the conversion increased with overall residence time when using oxygen; this increase in conversion was likely due to the higher concentration of oxygen present in the liquid phase. Notably, the use of oxygen decreased the selectivity by 25% compared to air. The observation of decreasing benzaldehyde selectivity with increasing benzyl alcohol conversion can be attributed to over-oxidation to benzyl benzoate and was consistent with previous reports for Pd catalysts [30].

Conversion in a three-phase packed bed reactor can be affected by fluid hydrodynamics, such as fluid flow regime, on catalyst wetting. Increasing the oxygen-to-reagent ratio in such reactors could improve conversion and selectivity [21]. Therefore, a tube-in-tube membrane contactor (sparger) was introduced as an alternative to the gas dosing module of the X-Cube™ in order to increase the oxygen-to-reagent ratio. The sparger contained a semipermeable inner tube constructed with Teflon AF-2400, which was highly gas-permeable to facilitate transport of oxygen from the gas into the liquid phase [31]. The sparger was configured such that the gas stream in the inner tube ran countercurrent to the liquid stream in the outer PTFE shell (Figure 2). This configuration was chosen based on modeling demonstrating a high saturation fraction [24]. The sparger was integrated in line with the existing X-Cube™ between the syringe and HPLC pumps. The liquid feed containing dissolved gas from the sparger was pumped directly into the X-Cube™.

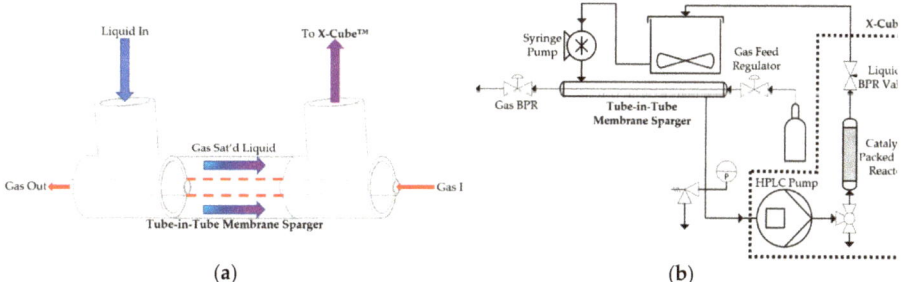

Figure 2. (a) Diagram of the tube-in-tube membrane sparger. The gas, red arrows, flows through the center tube counter-currently to the liquid fluid, blue arrows, in the shell. The gas diffuses through the membrane dissolving into the liquid, the blue arrows transition to purple to indicate a gas rich liquid stream. The gas rich liquid stream is fed to the packed bed. (b) Flow diagram of the continuous process with the external syringe pump and sparger added to the X-Cube™.

The integration of the sparger unit resulted in a significant increase in conversion and selectivity. The maximum conversion achieved after 10 passes was 10.9 ± 1.3% using air and 31.7 ± 6.1% using oxygen (Figure 3). At 2.8 bar, the sparger exhibited a 5-fold improvement using air and a 9-fold improvement using oxygen in conversion compared to the X-Cube™ gas dosing module with a fixed 1:19 volumetric gas to liquid ratio (dictated by the X-Cube™ system). The increase in conversion of benzyl alcohol indicated that the sparger improved the effective oxygen supplied to the catalyst surface compared to the gas dosing module. The maximum conversion of 31.7 ± 6.1% was consistent with a gas-limited system described by Yang and Jensen based on mass transfer modeling [24]. This result confirmed that the sparger serves to facilitate oxygen solubilization into the liquid stream according to Henry's law, which is a similar result to previous reports for industrially relevant gas/liquid reactions [25,26,28,32]. Conversion can be increased by diluting the substrate; however, dilution decreases the reactor productivity [10,26]. While conversion was an important metric for evaluating the efficiency of a chemical process, productivity/space time yield (STY) was an important consideration for reactor/process design. Reactor productivity/STY were especially relevant for continuous processes that incorporate recycling unreacted materials. Often, there exists a trade-off between selectivity/yield and productivity [33,34]. Practical examples include a time-dependent racemization process [35] and vigorously exothermic reductions [36].

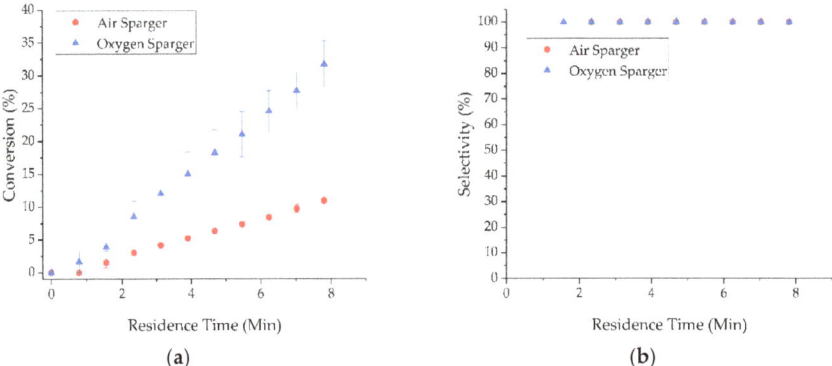

Figure 3. (a) Conversion of benzyl alcohol; (b) selectivity to benzaldehyde for the sparger modified X-Cube™ and external pump system for both air (red circles) and pure oxygen (blue triangles) at 2.8 bar.

It should be noted that 10% conversion was achieved in only 8 min total residence time, which equates to a STY of 22,000 g·L^{-1}·h^{-1} with air and 64,000 g·L^{-1}·h^{-1} with oxygen. Hii and co-workers previously reported achieving 100% conversion using a ruthenium based catalyst. To compare across different process conditions (2.5-fold higher pressure, ~50-fold lower benzyl alcohol concentration, different temperature, and different catalyst) we use STY. We note they only achieved a STY of approximately 1800 g·L^{-1}·h^{-1}, due in large part to the 50-fold lower benzyl alcohol concentration [19]. This work focused on productivity rather than obtaining 100% conversion; we were able to obtain a 10-fold higher STY using a different metal catalyst and increasing the substrate concentration at lower overall operating pressure compared to previous work [19]. Therefore, implementing the sparger in line with the X-Cube™ provided a straightforward approach to increasing reactor productivity at lower operating pressures from 25 bar [19] to 2.8 bar (this work).

Additionally, use of the sparger increased the reaction selectivity to 100% for both air and pure oxygen with no observable over-oxidation over ten passes (Figure 3). By comparison, over-oxidation was observed with the gas dosing module at the same pressure (Figure S2). The improvement in selectivity with the sparger was attributed to solubilization of the oxidant in the liquid phase. Specifically, segmented flow at a fixed 1:19 gas-liquid ratio resulted in poor gas-liquid mass transfer. Overall, this led to sequential presentation of the gas phase oxidant and then the reactant. This sequential presentation promoted overoxidation due to high concentration of oxygen on the catalyst surface [5,16,21,37]. In contrast with the segmented flow achieved by the X-Cube™ gas dosing module, the sparger solubilized the gas in the liquid phase before entering the packed bed, resulting in simultaneous presentation of the oxidant and reaction thus preventing overoxidation. This simultaneous presentation achieved a higher selectivity to benzaldehyde, even at lower pressures.

Next, the effect of pressure using the sparger was examined. The reaction pressure was increased to 9.7 bar and the conversion and selectivity were examined. At a residence time of four minutes, the conversion was increased from 5.2% to 11.7% with air (Figure 4a); however, the selectivity decreased to 84.8% (Figure 4b). The resulting STY was calculated to be 40,000 g·L^{-1}·h^{-1}. Similar results were observed using oxygen as well: conversion increased from 15.0% to 17.5% and selectivity decreased to 90% with a calculated STY of 54,000 g·L^{-1}·h^{-1} (Figure 4). These results suggested that at higher pressure, the resulting higher solubilized oxygen content enhanced reaction kinetics which improved conversion, but also promoted overoxidation which reduced selectivity. Therefore, operating pressure with the sparger was an important consideration to achieving selective oxidation.

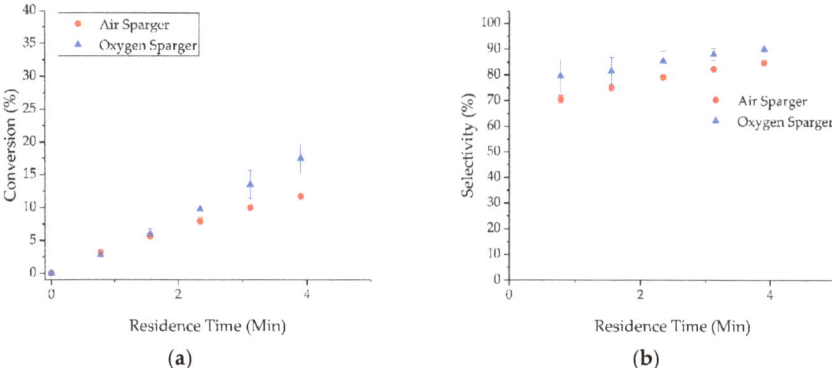

Figure 4. (a) Conversion of benzyl alcohol; (b) selectivity to benzaldehyde for the sparger modified X-Cube™ and external pump system for both air (red circles) and pure oxygen (blue triangles) at 9.7 bar.

The selectivity with the sparger using air at low (2.8 bar) pressure was approximately 10% better than the X-Cube™ at high (9.7 bar) pressures using oxygen as the oxidant (Table 1). Figure S3 contains the data for the X-Cube™ at high (9.7 bar) pressure. At low (2.8 bar) pressure, the productivity of the sparger with air increased 4-fold over the X-Cube™ with oxygen. Comparing the X-Cube™ at high (9.7 bar) pressure with the sparger at low (2.8 bar) pressure, there was a trade-off between conversion and selectivity indicated by the ~17% decrease in throughput and STY (Table 1). Importantly, using the sparger at lower pressure was beneficial because (1) lowering the operating pressure mitigated safety hazards and (2) improving the selectivity to 100% which reduced the complexity of separation. Overall, this example of a tube-in-tube membrane sparger integrated in-line with the X-Cube™ offered a simple alternative to the gas dosing module, with segmented flow, resulting in a significant increase in STY as well as improved selectivity at low (2.8 bar) pressure.

Table 1. Conversion, selectivity, throughput, and space time yield (STY) of the sparged process with air and non-sparged process with pure oxygen.

System	Pressure (Bar)	% Conversion	% Selectivity	Throughput (g/h)	STY (g/(L*h))
X-Cube™ O$_2$	9.7	17.5 ± 1.4	88.0 ± 1.8	60.3	77,000
Sparger O$_2$	2.8	31.7 ± 6.1	100 ± 0.0	49.7	64,000
	9.7	17.5 ± 3.8	90.1 ± 3.1	42.4	54,000
Sparger Air	2.8	10.9 ± 1.9	100 ± 0.0	17.1	22,000
	9.7	11.7 ± 0.5	84.8 ± 1.2	31.1	40,000

3. Materials and Methods

Benzyl alcohol (99% pure) and triethylene glycol dimethyl ether (99% pure) were received from Acros Organics. Toluene (99.9% pure) and palladium on activated carbon (10 wt% loading) were received from Sigma-Aldrich (St. Louis, MO, USA) Oxygen (Ox R, 99.999% pure) and air (Ultra Zero, 21% pure) were supplied by Praxair (Danbury, CT, USA). For lower pressure reactions, house air was used. All chemicals were used as received from commercial sources without any purification.

A commercially-available reactor, the X-Cube™ (ThalesNano, Budapest, Hungary), was employed to perform the reactions. An external reciprocating syringe pump (Chemtrix, Echt, The Netherlands) was added in line as the primary pump in order to maintain consistent liquid flow. A catalyst cartridge, packed with 100 mg of catalyst, was installed into the X-Cube™ mantle. Typically, the pump inlet and the reactor outlet lines were placed in a 40 mL solution of 50 vol.% benzyl alcohol (4.8 M) and 10 vol.% triglyme (0.55 M, the internal standard) in toluene. Oxygen or air was introduced to the reaction solution before entering the packed bed. The liquid flow rate into the system was 1 mL/min and the reactor was set to 100 °C. The liquid back pressure was maintained by the X-Cube™ BPR valve. The solution was then passed 10 times; each pass was characterized by gas chromatography (FID detector, HP-5MS column, 6890N, Agilent, Santa Clara, CA, USA) and the identity of the constituents in solution were confirmed with GC-mass spectroscopy (MS Detector 5973, HP-5MS column, 6890N, Agilent, Santa Clara, CA, USA).

Gas introduction was controlled using the gas dosing module in the X-Cube™ or a tube-in-tube membrane sparger depending on the experiment. The gas dosing module was set to maintain a 1:19 gas bubble to liquid ratio within the reactor. The membrane sparger was developed to increase the oxygen available for reaction and integrated with the existing system (Figure 2), replacing the gas dosing module. The sparger was configured with gas in the inner tube (Teflon AF-2400, Biogeneral, San Diego, CA, USA) flowing counter-currently to the liquid stream in the outer shell (PTFE, Restek, Bellefonte, PA, USA). The inner tubing had an OD and ID of 1.0 and 0.8 mm and the outer tubing had an OD and ID of 0.125 and 0.063 inches, respectively. The tubing was joined by two 1/8-inch tee joints (Swagelok, Solon, OH, USA) and the inner tubing was fitted into special 1/16" tubing with a bore of 0.040" (VICI, Houston, TX, USA). The gas flow rate, measured at 1227 mL/min, was controlled by a pressure differential between the cylinder regulator (Praxair, Danbury, CT, USA), and a back pressure regulator (Swagelok, Solon, OH, USA) at the outlet of the inner tube. A pressure relief valve (Swagelok, Solon, OH, USA) was added to prevent over pressurization of the sparger. The liquid from the sparger was fed to the X-Cube™ and the oxidation reaction and characterization was carried out as described above.

The system was characterized by benzyl alcohol conversion and selectivity. Conversion (X_i) was based on the consumption of benzyl alcohol and was calculated according to Equation (1):

$$X_i = \frac{N_{alcohol,\,in} - N_{alcohol,out}}{N_{alcohol,in}} * 100\% \tag{1}$$

where $N_{alcohol,in}$ and $N_{alcohol,out}$ were the mols of benzyl alcohol at the inlet and outlet, respectively. The selectivity (S) of the desired product, benzaldehyde, relative to the undesired side products, typically benzyl benzoate, was calculated according to Equation (2):

$$S = \frac{N_{aldehyde}}{N_{products}} * 100\% \tag{2}$$

where $N_{aldehyde}$ was the number of mols of benzaldehyde produced and $N_{products}$ was the total moles of all products. The amount of product produced, throughput (TH, g/h), was calculated based on Equation (3):

$$TH = \frac{N_{aldehyde} * MW_{aldehyde}}{t_{rxn}} \tag{3}$$

where $MW_{aldehyde}$ represented the molecular weight of benzaldehyde and t_{rxn} was the overall reaction time in hours. Space-time yield (STY, g/(L*h)) was calculated by Equation (4):

$$STY = \frac{N_{aldehyde} * MW_{aldehyde}}{V_r * t_{rxn}} \tag{4}$$

where V_r is the packed bed reactor volume (0.78 mL, CatCart 4 mm × 62 mm).

4. Conclusions

We have investigated a simple approach to improve multiphase catalytic reactions performed in a flow chemistry unit. Specifically, we integrated a tube-in-tube membrane contactor (sparger) to the X-Cube™ reactor and examined oxidation of benzyl alcohol using a heterogeneous palladium catalyst. The sparger increased the conversion 9-fold using oxygen and 5-fold using air as the oxidant compared to the gas-dosing module of the X-Cube™ at low (2.8 bar) pressure operation. Furthermore, the sparger increased the reaction selectivity of benzaldehyde up to 100% compared to ~60–80% without the sparger. The sparger facilitated these improvements by enabling the simultaneous addition of the dissolved gas oxidant and the reagents to the catalyst in the same phase through facilitated gas solubilization. The sparger also allowed for safer operation of these gas-liquid processes since the sparged process was performed at lower pressure and achieved similar results, i.e. throughput within 20%, compared to the segmented flow process at higher pressures. Thus, sparger implementation offered opportunities to improve process safety while also improving reactor productivity as the space time yield increased by 4-fold using air with the sparger compared to oxygen with the X-Cube™ at 2.8 bar operation. Overall, we anticipated this approach can be broadly applied to other multiphase catalytic reactions.

Supplementary Materials: The following are available online at http://www.mdpi.com/2073-4344/9/1/95/s1, Figure S1: The schematic diagram of the X-Cube™ reactor's components (provided by ThalesNano), Figure S2: (**a**) conversion of benzyl alcohol; (**b**) selectivity to benzaldehyde for the X-Cube™ and external pump system for both air (red circles) and pure oxygen (blue triangles) at 2.8 bar, Figure S3: (**a**) Conversion of benzyl alcohol; (**b**) selectivity to benzaldehyde for the X-Cube™ and external pump system for both air (red circles) and pure oxygen (blue triangles) at 9.7 bar.

Author Contributions: Conceptualization, M.B., S.E.G.III, C.T. and B.F.G.; Investigation, M.B. and A.L.; Validation, M.B., S.E.G.III, A.L., C.T. and B.F.G.; Formal Analysis, M.B. and C.T.; Resources, C.T. and B.F.G.; Writing—Original Draft Preparation, M.B. and S.G.; Writing—Review & Editing, C.T. and B.F.G.; Visualization, M.B. and C.T.; Supervision, C.T. and B.F.G.; Project Administration, C.T. and B.F.G.; Funding Acquisition, C.T. and B.F.G.

Funding: This research was partially supported by VCU College of Engineering and the Center for Rational Catalyst Synthesis an Industry/University Cooperative Research Center funded in part by the National Science Foundation (Industry/University Collaborative Research Center IIP 1464630 grant ID 17-3171).

Acknowledgments: The financial support of the VCU College of Engineering and National Science Foundation, USA (IIP 1464630 grant ID 17-3171) is gratefully acknowledged. The authors thank Evan Pfab for his assistance with experiments.

Conflicts of Interest: The authors declare no conflicts of interest.

References

1. Pieber, B.; Kappe, C.O. Aerobic Oxidations in Continuous Flow. *Top. Organomet. Chem.* **2016**, *57*, 97–136. [CrossRef]
2. Gavriilidis, A.; Constantinou, A.; Hellgardt, K.; Hii, K.K.; Hutchings, G.J.; Brett, G.L.; Kuhn, S.; Marsden, S.P. Aerobic oxidations in flow: Opportunities for the fine chemicals and pharmaceuticals industries. *React. Chem. Eng.* **2016**, *1*, 595–612. [CrossRef]
3. Mills, P.L.; Quiram, D.J.; Ryley, J.F. Microreactor technology and process miniaturization for catalytic reactions-A perspective on recent developments and emerging technologies. *Chem. Eng. Sci.* **2007**, *62*, 6992–7010. [CrossRef]
4. Losey, M.W.; Schmidt, M.A.; Jensen, K.F. Microfabricated multiphase packed-bed reactors: Characterization of mass transfer and reactions. *Ind. Eng. Chem. Res.* **2001**, *40*, 2555–2562. [CrossRef]
5. Hessel, V.; Angeli, P.; Gavriilidis, A.; Löwe, H. Gas-liquid and gas-liquid-solid microstructured reactors: Contacting principles and applications. *Ind. Eng. Chem. Res.* **2005**, *44*, 9750–9769. [CrossRef]
6. Plutschack, M.B.; Pieber, B.; Gilmore, K.; Seeberger, P.H. The Hitchhiker's Guide to Flow Chemistry. *Chem. Rev.* **2017**, *117*, 11796–11893. [CrossRef] [PubMed]
7. Geyer, K.; Codée, J.D.C.; Seeberger, P.H. Microreactors as tools for synthetic Chemists—The chemists' round-bottomed flask of the 21st century? *Chem. Eur. J.* **2006**, *12*, 8434–8442. [CrossRef]

8. Gemoets, H.P.L.; Su, Y.; Shang, M.; Hessel, V.; Luque, R.; Noël, T. Liquid phase oxidation chemistry in continuous-flow microreactors. *Chem. Soc. Rev.* **2016**, *45*, 83–117. [CrossRef] [PubMed]
9. Rossetti, I.; Compagnoni, M. Chemical reaction engineering, process design and scale-up issues at the frontier of synthesis: Flow chemistry. *Chem. Eng. J.* **2016**, *296*, 56–70. [CrossRef]
10. Jensen, K.F.; Reizman, B.J.; Newman, S.G. Tools for chemical synthesis in microsystems. *Lab Chip* **2014**, *14*, 3206–3212. [CrossRef] [PubMed]
11. Caron, S.; Dugger, R.W.; Ruggeri, S.G.; Ragan, J.A.; Brown Ripin, D.H. Large-scale oxidations in the pharmaceutical industry. *Chem. Rev.* **2006**, *106*, 2943–2989. [CrossRef] [PubMed]
12. Zotova, N.; Roberts, F.J.; Kelsall, G.H.; Jessiman, A.S.; Hellgardt, K.; Hii, K.K. (Mimi) Catalysis in flow: Au-catalysed alkylation of amines by alcohols. *Green Chem.* **2012**, *14*, 226–232. [CrossRef]
13. Dijksman, A.; Marino-González, A.; Mairata i Payeras, A.; Arends, I.W.C.E.; Sheldon, R.A. Efficient and Selective Aerobic Oxidation of Alcohols into Aldehydes and Ketones Using Ruthenium/TEMPO as the Catalytic System. *J. Am. Chem. Soc.* **2001**, *123*, 6826–6833. [CrossRef] [PubMed]
14. Nowicka, E.; Hofmann, J.P.; Parker, S.F.; Sankar, M.; Lari, G.M.; Kondrat, S.A.; Knight, D.W.; Bethell, D.; Weckhuysen, B.M.; Hutchings, G.J. In situ spectroscopic investigation of oxidative dehydrogenation and disproportionation of benzyl alcohol. *Phys. Chem. Chem. Phys.* **2013**, *15*, 12147. [CrossRef] [PubMed]
15. Zhou, C.; Chen, Y.; Guo, Z.; Wang, X.; Yang, Y. Promoted aerobic oxidation of benzyl alcohol on CNT supported platinum by iron oxide. *Chem. Commun.* **2011**, *47*, 7473. [CrossRef] [PubMed]
16. Williams, R.; Medlin, J. Benzyl alcohol oxidation on Pd (111): Aromatic binding effects on alcohol reactivity. *Langmuir* **2014**, *30*, 4642–4653. [CrossRef] [PubMed]
17. Besson, M.; Gallezot, P. Selective oxidation of alcohols and aldehydes on metal catalysts. *Catal. Today* **2000**, *57*, 127–141. [CrossRef]
18. Albonetti, S.; Mazzoni, R.; Cavani, F. *Homogeneous, Heterogeneous and Nanocatalysis*; The Royal society of Chemistry: London, UK, 2015; ISBN 9781782621652.
19. Zotova, N.; Hellgardt, K.; Kelsall, G.H.; Jessiman, A.S.; Hii, K.K. (Mimi) Catalysis in flow: The practical and selective aerobic oxidation of alcohols to aldehydes and ketones. *Green Chem.* **2010**, *12*, 2157. [CrossRef]
20. Miedziak, P.; Sankar, M.; Dimitratos, N.; Lopez-Sanchez, J.A.; Carley, A.F.; Knight, D.W.; Taylor, S.H.; Kiely, C.J.; Hutchings, G.J. Oxidation of benzyl alcohol using supported gold-palladium nanoparticles. *Catal. Today* **2011**, *164*, 315–319. [CrossRef]
21. Al-Rifai, N.; Galvanin, F.; Morad, M.; Cao, E.; Cattaneo, S.; Sankar, M.; Dua, V.; Hutchings, G.; Gavriilidis, A. Hydrodynamic effects on three phase micro-packed bed reactor performance—Gold-palladium catalysed benzyl alcohol oxidation. *Chem. Eng. Sci.* **2016**, *149*, 129–142. [CrossRef]
22. Wu, G.; Constantinou, A.; Cao, E.; Kuhn, S.; Morad, M.; Sankar, M.; Bethell, D.; Hutchings, G.J.; Gavriilidis, A. Continuous heterogeneously catalyzed oxidation of benzyl alcohol using a tube-in-tube membrane microreactor. *Ind. Eng. Chem. Res.* **2015**, *54*, 4183–4189. [CrossRef]
23. O'Brien, M.; Taylor, N.; Polyzos, A.; Baxendale, I.R.; Ley, S.V. Hydrogenation in flow: Homogeneous and heterogeneous catalysis using Teflon AF-2400 to effect gas–liquid contact at elevated pressure. *Chem. Sci.* **2011**, *2*, 1250–1257. [CrossRef]
24. Yang, L.; Jensen, K.F. Mass transport and reactions in the tube-in-tube reactor. *Org. Process Res. Dev.* **2013**, *17*, 927–933. [CrossRef]
25. Wu, G.; Cao, E.; Kuhn, S.; Gavriilidis, A. A Novel Approach for Measuring Gas Solubility in Liquids Using a Tube-in-Tube Membrane Contactor. *Chem. Eng. Technol.* **2017**, 1–6. [CrossRef]
26. Petersen, T.P.; Polyzos, A.; O'Brien, M.; Ulven, T.; Baxendale, I.R.; Ley, S.V. The oxygen-mediated synthesis of 1,3-butadiynes in continuous flow: Using teflon AF-2400 to effect gas/liquid contact. *ChemSusChem* **2012**, *5*, 274–277. [CrossRef] [PubMed]
27. Constantinou, A.; Wu, G.; Corredera, A.; Ellis, P.; Bethell, D.; Hutchings, G.J.; Kuhn, S.; Gavriilidis, A. Continuous Heterogeneously Catalyzed Oxidation of Benzyl Alcohol in a Ceramic Membrane Packed-Bed Reactor. *Org. Process Res. Dev.* **2015**, *19*, 1973–1979. [CrossRef]
28. Skowerski, K.; Czarnocki, S.J.; Knapkiewicz, P. Tube-in-tube reactor as a useful tool for homo- and heterogeneous olefin metathesis under continuous flow mode. *ChemSusChem* **2014**, *7*, 536–542. [CrossRef]
29. Brzozowski, M.; O'Brien, M.; Ley, S.V.; Polyzos, A. Flow chemistry: Intelligent processing of gas-liquid transformations using a tube-in-tube reactor. *Acc. Chem. Res.* **2015**, *48*, 349–362. [CrossRef]

30. Feng, J.; Ma, C.; Miedziak, P.J.; Edwards, J.K.; Brett, G.L.; Li, D.; Du, Y.; Morgan, D.J.; Hutchings, G.J. Au–Pd nanoalloys supported on Mg–Al mixed metal oxides as a multifunctional catalyst for solvent-free oxidation of benzyl alcohol. *Dalton Trans.* **2013**, *42*, 14498–14508. [CrossRef]
31. Greene, J.F.; Preger, Y.; Stahl, S.S.; Root, T.W. PTFE-Membrane Flow Reactor for Aerobic Oxidation Reactions and Its Application to Alcohol Oxidation. *Org. Process Res. Dev.* **2015**, *19*, 858–864. [CrossRef]
32. Koos, P.; Gross, U.; Polyzos, A.; O'Brien, M.; Baxendale, I.; Ley, S.V. Teflon AF-2400 mediated gas–liquid contact in continuous flow methoxycarbonylations and in-line FTIR measurement of CO concentration. *Org. Biomol. Chem.* **2011**, *9*, 6903–6908. [CrossRef] [PubMed]
33. Schweidtmann, A.M.; Clayton, A.D.; Holmes, N.; Bradford, E.; Bourne, R.A.; Lapkin, A.A. Machine learning meets continuous flow chemistry: Automated optimization towards the Pareto front of multiple objectives. *Chem. Eng. J.* **2018**, *352*, 277–282. [CrossRef]
34. Jumbam, D.N.; Skilton, R.A.; Parrott, A.J.; Bourne, R.A.; Poliakoff, M. The Effect of Self-Optimisation Targets on the Meethylation of Alcohols Using Dimethyl Carbonate in Supercritical CO_2. *J. Flow Chem.* **2012**, *2*, 24–27. [CrossRef]
35. Erdmann, V.; Mackfeld, U.; Rother, D.; Jakoblinnert, A. Enantioselective, continuous (R)- and (S)-2-butanol synthesis: Ahieving high space-time yields with recombinant E. coli cells in a micro-aqueous, solvent-free reaction system. *J. Biotechnol.* **2014**, *191*, 106–112. [CrossRef] [PubMed]
36. Schotten, C.; Howard, J.L.; Jenkins, R.L.; Codina, A.; Browne, D.L. A continuous flow-batch hybrid reactor for commodity chemical synthesis enabled by inline NMR and temperature monitoring. *Tetrahedon* **2018**, *74*, 5503–5509. [CrossRef]
37. Rodríguez-Reyes, J.C.F.; Friend, C.M.; Madix, R.J. Origin of the selectivity in the gold-mediated oxidation of benzyl alcohol. *Surf. Sci.* **2012**, *606*, 1129–1134. [CrossRef]

© 2019 by the authors. Licensee MDPI, Basel, Switzerland. This article is an open access article distributed under the terms and conditions of the Creative Commons Attribution (CC BY) license (http://creativecommons.org/licenses/by/4.0/).

Article

Catalytic Hydrodechlorination of Chlorophenols in a Continuous Flow Pd/CNT-Ni Foam Micro Reactor Using Formic Acid as a Hydrogen Source

Jun Xiong *,† and Ying Ma †

School of Pharmacy, Zunyi Medical University, Guizhou 510640, China; maying160@163.com
* Correspondence: xiongjun@zmc.edu.cn; Tel.: +86-0851-2864-2341
† These authors contributed equally to this work.

Received: 16 December 2018; Accepted: 8 January 2019; Published: 12 January 2019

Abstract: Catalytic hydrodechlorination (HDC) has been considered as a promising method for the treatment of wastewater containing chlorinated organic pollutants. A continuous flow Pd/carbon nanotube (CNT)-Ni foam micro reactor system was first developed for the rapid and highly efficient HDC with formic acid (FA) as a hydrogen source. This micro reactor system, exhibiting a higher catalytic activity of HDC than the conventional packed bed reactor, reduced the residence time and formic acid consumption significantly. The desired outcomes (dichlorination >99.9%, 4-chlorophenol outlet concentration <0.1 mg/L) can be obtained under a very low FA/substrate molar ratio (5:1) and short reaction cycle (3 min). Field emission scanning electron microcopy (FESEM) and deactivation experiment results indicated that the accumulation of phenol (the main product during the HDC of chlorophenols) on the Pd catalyst surface can be the main factor for the long-term deactivation of the Pd/CNT-Ni foam micro reactor. The catalytic activity deactivation of the micro reactor could be almost completely regenerated by the efficient removal of the absorbed phenol from the Pd catalyst surface.

Keywords: catalytic hydrodechlorination; micro reactor; chlorophenols; Pd catalyst

1. Introduction

Chlorophenols (CPs), existing widely in polluted groundwater and the wastewater effluents of industry [1,2], have been listed as priority pollutants in many countries because of their high toxicity, adverse environmental impacts and poor biodegradability [3–5]. For the safe disposal of these highly toxic chlorinated organic pollutants, many detoxification techniques such as biodegradation [6], photochemical degradation [5], advanced oxidation [7,8] and catalytic hydrodechlorination [9,10] have been proposed. Among the available water treatment techniques mentioned above, hydrodechlorination (HDC) presents the advantages of greater flexibility, low energy consumption and relatively safe by-products, showing promising prospects in the treatment of wastewater containing chlorinated organic pollutants [11–13].

Molecular hydrogen (H_2) is the most widely employed hydrogen source for HDC [10,14–18]. However, the low H_2 utilization efficiency caused by its very low solubility in water (0.84 mM, at 288 K and P_{H_2} = 100 kPa) [19,20], and the low process safety associated with hydrogen gas usage [21,22], may lead to some adverse influences during the treatment of wastewater on a large scale through HDC. To overcome these drawbacks, other potential hydrogen sources, including isopropanol, hydrazine, formic acid and formate, have been reported [19,20,23,24]. Formic acid, with high solubility in water, has been proven to be a promising alternative hydrogen source of HDC [14,20].

The HDC of chlorinated organic compounds are generally performed in a conventional batch reactor with a powder Pd catalyst [11,25–28]. A high formic acid (FA) consumption (FA/substrate molar ratio ≥35:1) and relatively long reaction time (tens of minutes or several hours) is required for a high dechlorination of the substrate during HDC, as FA is utilized to provide hydrogen [14,19,20,23]. Furthermore, the inherent difficulty of scale-up is still a challenge during the industrial application of a batch reactor system [29,30]. The application of a small-scale continuous flow reactor (micro reactor) system may offer some advantages, especially for heterogeneous catalytic reactions such as HDC, over a conventional batch reactor system [29]. The micro reactor, with short diffusion paths and large interfacial areas, can provide efficient mixing and fast mass transfer during HDC, contributing to the achievement of desired outcomes (e.g., high dechlorination, low FA consumption and short reaction cycle) [30–32]. Scaling up a continuous reactor system is also generally easier than scaling up a batch reactor system [30,33,34]. Thus, it is interesting to develop a new micro reactor system for HDC in continuous flow.

A carbon nanotube (CNT)-Ni foam-supported Pd catalyst has been found to be an excellent catalyst candidate for HDC [35]. Notably, this monolithic Pd/CNT-Ni foam composite catalyst can also be utilized as a micro reactor, due to its highly porous channel with a micrometer size and well-dispersed active Pd catalyst on the surface of the micro channel. Herein, we report the usage of a micro reactor system for conducting HDC in continuous flow for the first time. Our aim is to develop a new micro reactor system possessing a high catalytic HDC performance. The Pd/CNT-Ni foam micro reactor system was configured and its HDC of CPs was evaluated in detail by using the safe and efficient formic acid as a hydrogen source. Finally, the long-term deactivation and regeneration of the Pd/CNT-Ni foam micro reactor were investigated.

2. Results and Discussion

2.1. Catalyst Characterization

Figure 1 presents Field emission scanning electron microcopy (FESEM) results of the Pd/CNT-Ni foam micro reactor. A highly porous Ni foam skeleton with irregular micro channels of about 150–500 μm in size (Figure 1a) was found in this monolithic micro reactor. Uniformly dispersed Pd nanoparticles were observed on the CNT surface of the fresh Pd/CNT-Ni foam micro reactor, and their amount increased significantly with the Pd loading (Figure 1b,c). The micro reactors were used for 76 h and deactivated with a high-concentration phenol solution, following which similar organic layers were observed on the Pd catalyst surface (Figure 1d,e). After regeneration treatment, the organic layer covered on the Pd catalyst surface disappeared (Figure 1f), indicating an efficient removal of the absorbed organic layer from the catalyst surface during regeneration.

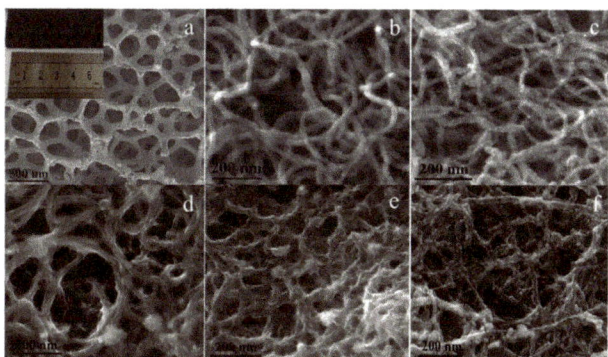

Figure 1. Field emission scanning electron microcopy (FESEM) images of 0.5 wt.% fresh (**a**,**c**), 0.2 wt.% fresh (**b**), 0.5 wt.% used (**d**), 0.5 wt.% deactivated with phenol (**e**) and 0.5 wt.% regenerated (**f**) Pd/carbon nanotube (CNT)-Ni foam micro reactors. The inset in (**a**) is the photograph of the Pd/CNT-Ni foam micro reactor.

The fresh and used Pd/CNT-Ni foam micro reactors were also investigated by TEM/HRTEM. Figure 2a shows that the Pd particles were well dispersed in a fresh Pd/CNT-Ni foam micro reactor, which is consistent with the FESEM observation. The average particle size was about 2.68 nm (Figure 2c). The nanoparticle composition was confirmed to be metal Pd by the HRTEM results and the corresponding fast Fourier transform (FFT). The lattice plane with the interplanar distance of 2.26 Å and its corresponding FFT pattern, assigned to the (111) plane of the face-centered cubic (fcc) Pd, are presented in Figure 2e [36,37]. The Pd particle characteristics including dispersion, size distribution and composition in the used micro reactor (Figure 2b,d,f) are almost the same as those presented in the fresh micro reactor, revealing the excellent stability of Pd particles during HDC.

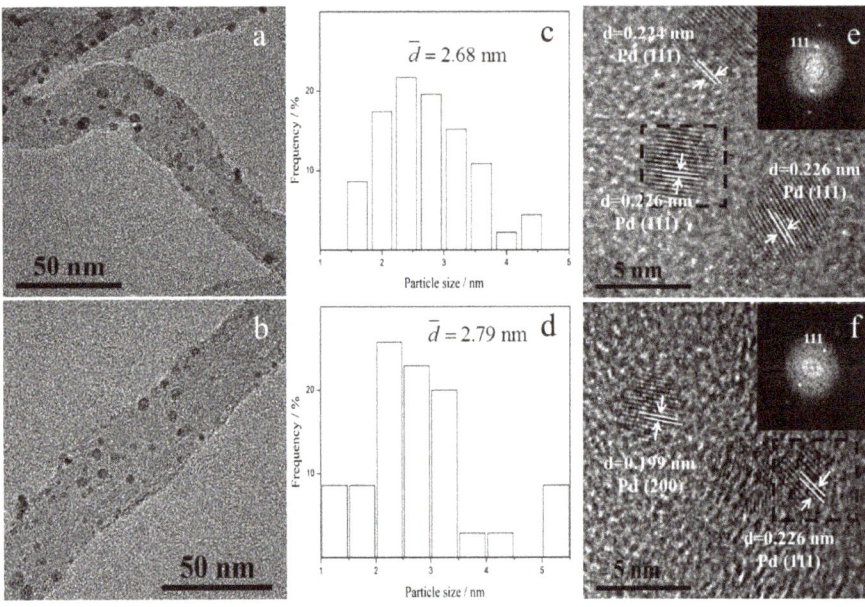

Figure 2. TEM (**a**,**b**) images and HRTEM (**e**,**f**) images of the fresh and used Pd/CNT-Ni foam micro reactors, respectively. (**c**,**d**) The corresponding particle size distribution of (**a**,**b**). The insets in (**e**,**f**) are the fast Fourier transform (FFT) patterns corresponding to the labeled regions.

2.2. HDC of 4-CP in Continuous Flow Micro Reactor and Packed Bed Reactor

For a comparison of the catalytic performance between the micro reactor and packed bed reactor, the HDC evaluations were controlled at analogous reaction conditions, such as the same reactor volume (about 4.8 mL), residence time and FA/substrate molar ratio. As shown in Figure 3, a significant increase in dechlorination and decrease in 4-CP outlet concentration was observed when switching from the packed bed reactor to the micro reactor. Desired outcomes (dichlorination >99.9%, 4-CP outlet concentration <0.1 mg/L) of HDC was obtained in the micro reactor. With the reduction of the FA/substrate molar ratio from 5:1 to 2:1, the enhancement of catalytic activity in the micro reactor would be increased.

For HDC with the alternative hydrogen source of FA (a typical liquid–solid heterogeneous catalytic process), the reaction rate depends on the concentration of the reactants (4-CP and FA) available on the catalyst surface sites, which means that the observed catalytic activity is influenced by both the surface hydrodechlorination reaction rate and the reactants' mass transport rate from the liquid phase to the catalyst surface [34,38]. In this work, the reaction scale is reasonable enough to be the main difference between the micro reactor and packed bed reactor, because these two heterogeneous catalytic reactors possess the same catalyst characteristics, including active Pd loading (0.5 wt.%) and Pd particle size (about 2.68 nm). Generally, the reduction of the reaction channel to a micro size is beneficial for enhancing the mass transport rate of reactants significantly [29,31,32,39]. Therefore, a higher HDC activity was exhibited by the micro reactor. In HDC, by using FA as a hydrogen source, a higher FA concentration on catalyst surface sites is necessary to generate a sufficient Pd-H reactive species, due to the incomplete FA decomposition during reaction [14,23]. A micro reactor with a very short diffusion distance can undoubtedly improve the refill of FA on the catalyst surface, especially under a low FA/substrate molar ratio. That is why the more obvious advantage of the micro reactor was observed under a lower FA/substrate molar ratio.

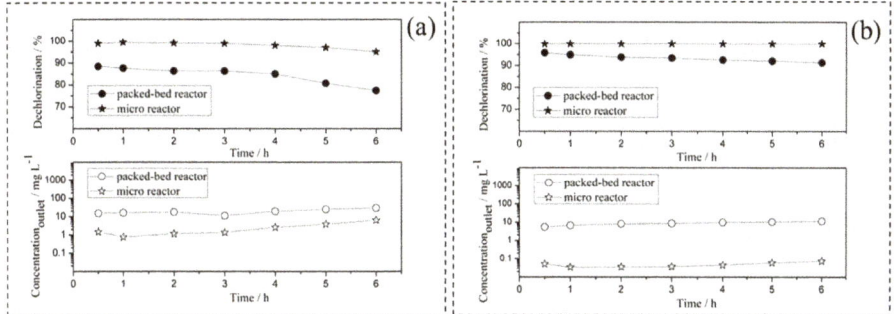

Figure 3. Hydrodechlorination (HDC) performance under 2:1 formic acid (FA)/substrate molar ratio (**a**) and 5:1 FA/substrate molar ratio (**b**) in different continuous flow reactors. Reaction conditions: residence time (3 min), Pd loading (0.5 wt.%), 4-chlorophenol concentration (1 mmol/L).

2.3. HDC of 4-CP in Micro Reactor with Different Pd Contents

Figure 4 shows 4-CP dechlorination in a continuous flow Pd/CNT-Ni foam micro reactor with different Pd contents. For the micro reactor without Pd, a very low dechlorination (<5%) of 4-CP was found. A significant decrease of 4-CP dechlorination from 80.2% to 47.8% during the first 6 h was presented by the micro reactor with 0.1 wt.% Pd. The catalytic activity and stability were obviously enhanced by the increase of Pd loading from 0.1 to 0.5 wt.%. A very highly efficient and stable HDC of 4-CP was exhibited by the micro reactor with 0.5 wt.% Pd. No significant changes of 4-CP dechlorination and catalytic stability were observed with the further increase of Pd content in the micro reactor from 0.5 to 2 wt.%.

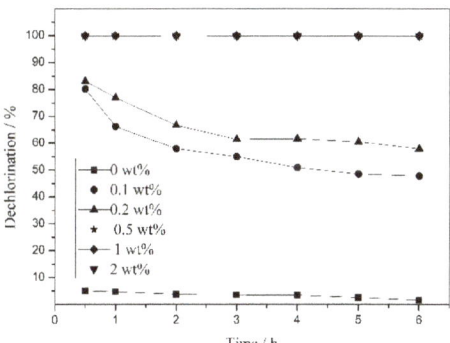

Figure 4. Dechlorination of 4-CP during an HDC reaction in a Pd/CNT-Ni foam micro reactor with different Pd loadings. Reaction conditions: residence time (3 min), FA/substrate molar ratio (5:1), 4-chlorophenol concentration (1 mmol/L).

In the HDC mechanism (Scheme 1) [38], the CPs (4-CP) were adsorbed on the surface sites of the Pd catalyst and consumed by the surface hydrodechlorination reaction, involving the adsorbed atomic hydrogen and adsorbed CPs. The accumulation of 4-CP on the catalyst surface may occur, as the surface reaction rate is very low. For the micro reactors with a lower Pd loading (less than 0.2 wt.%), the surface reaction rate is expected to be lower, due to the limited availability of active Pd sites (Figure 1b). Jadbabaei et al. [38] reported the obvious catalytic deactivation caused by the 4-CP accumulation on the surface of a catalyst with a low Pd loading (<0.44 wt.%) during the HDC of 4-CP. This phenomenon was also observed during our study. For the micro reactor with 0.1 wt.% Pd, the 4-CP dechlorination decreased rapidly from 80.2% to 47.8%. The catalyst active surface sites in the low Pd loading micro reactor were gradually covered by the accumulating 4-CP during the reaction, resulting in the reduction of the HDC catalytic activity. Thus, the micro reactors with lower Pd loadings (0.1 and 0.2 wt.%) present lower catalytic activity and stability. As the Pd loading in the micro reactor increased, the surface reaction could be accelerated by the increase in active Pd sites (Figure 1b,c) [40,41]. Additionally, the accumulation of 4-CP on the catalyst surface was alleviated or even eliminated. Therefore, high catalytic activity (dechlorination >99.9%) and stability was exhibited by the micro reactor with a relatively high Pd loading (≥0.5 wt.%).

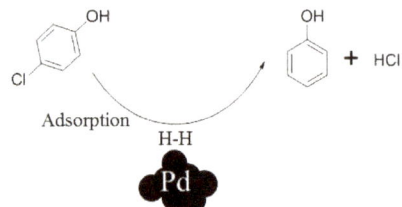

Scheme 1. Mechanism of HDC on the Pd catalyst surface.

2.4. HDC of 4-CP in Micro Reactor under Different FA/Substrate Molar Ratios

The FA decomposition rate is considered to be slower than the hydrodechlorination rate during HDC, although it can be stimulated by hydrodechlorination simultaneously [14,20]. A high FA/substrate molar ratio is required to provide sufficient active adsorbed atomic hydrogen for a highly efficient HDC process. The influence of the FA/substrate molar ratio on the dechlorination of 4-CP during HDC in continuous flow Pd/CNT-Ni foam micro reactor was evaluated. As shown in Figure 5, the catalytic activity and stability were enhanced significantly with the increase of the FA/substrate molar ratio from 1:1 to 5:1. The lower activity and stability under low FA/substrate molar ratios (1:1 and 2:1) may also be attributed to the low 4-CP consumption rate on the catalyst surface caused by the insufficiently active adsorbed atomic hydrogen under these conditions [19,38]. Notably, the FA/substrate molar ratio required for the almost complete dechlorination of 4-CP during HDC in our study was far less than that reported in other works [20,23].

2.5. The Longevity, Deactivation and Regeneration of Pd/CNT-Ni Foam Micro Reactor

An HDC durability test was conducted for the evaluation of long-term catalytic stability in a continuous flow Pd/CNT-Ni foam micro reactor. Figure 6a shows the 4-CP dechlorination as a function of reaction time. The 4-CP dechlorination maintains a very high and stable value (≥99.9%) during the initial 30 h. However, a gradual loss of catalytic activity with reaction time was observed after 30 h. Only 60.9% dechlorination of 4-CP was found at the final 76th hour. The long-term deactivation of HDC could occur in a continuous flow Pd/CNT-Ni foam micro reactor.

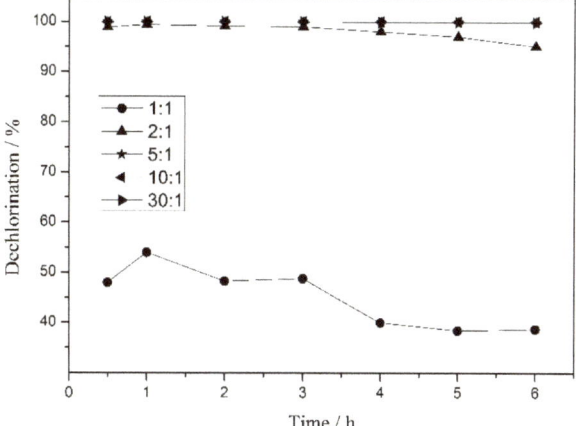

Figure 5. Dechlorination of 4-CP during HDC reaction in a Pd/CNT-Ni foam micro reactor under different FA/substrate molar ratios. Reaction conditions: residence time (3 min), Pd loading (0.5 wt.%), 4-chlorophenol concentration (1 mmol/L).

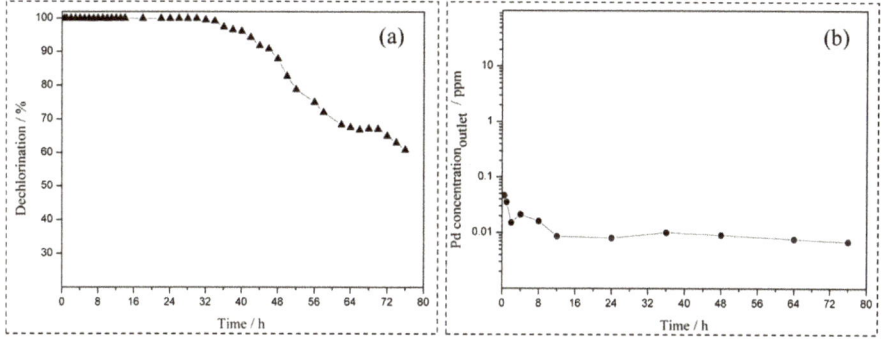

Figure 6. Dechlorination of 4-CP (**a**) and Pd outlet concentration (**b**) as a function of reaction time during HDC reaction in a Pd/CNT-Ni foam micro reactor. Reaction conditions: residence time (3 min), Pd loading (0.5 wt.%), FA/substrate molar ratio (5:1), 4-chlorophenol concentration (1 mmol/L).

HCl poisoning, which may induce severe Pd leaching and the alteration of Pd particles through corrosive action, has been widely reported as an important factor of Pd catalyst deactivation during HDC [12,38,42,43]. In our study, the continuous flow operation in the micro reactor achieved a more effective removal of HCl from the Pd catalyst surface, limiting the influences of HCl by-products during HDC [44]. Additionally, Pd particles with excellent stability, which may result from the strong interaction between the transition metal (Pd) atoms and sp2-hybridized carbon atoms [12,45], is expected to be resistant to HCl poisoning during the very short residence time (3 min). These inferences were supported by the Pd leaching measurement and TEM/HRTEM analysis. As illustrated in Figure 6b, Pd leaching in the effluents of the micro reactor was very low (<0.05 ppm) and almost undetectable (<0.01 ppm) after the 10th hour. The dispersion, distribution and composition of Pd particles were also largely unchanged during 76 hours of reaction time (Figure 2). Therefore, the influence of HCl poisoning on the catalytic stability was not significant during HDC in the Pd/CNT-Ni foam micro reactor.

The fresh and used Pd/CNT-Ni foam micro reactors were further investigated by FESEM. It was found that the Pd catalyst surface in the Pd/CNT-Ni foam micro reactors was covered by an organic layer after 76 h (Figure 1c,d). The used Pd/CNT-Ni foam micro reactor was soaked in ethanol for 4 h to perform an organics partial desorption test. Many phenols (about 37.8 mg/L in solution), the main product for the HDC of 4-CP, were detected. To confirm the role of phenol accumulation on the deactivation of the Pd/CNT-Ni foam micro reactor, a deactivation experiment with a high-concentration phenol solution (1 mol/L) treatment was carried out for 4 h. A similar organic layer to that caused by phenol accumulation was observed on the catalyst surface of the deactivated micro reactor (Figure 1e). According to the HDC performance results shown in Figure 7, serious deactivation of the Pd/CNT-Ni foam micro reactor was observed. The 4-CP dechlorination decreased remarkably from almost 100% to a very low level, less than 15%. Serious deactivation of HDC induced by phenol accumulation on the Pd catalyst surface was also observed in the work of Jadbabaei et. al. [38].

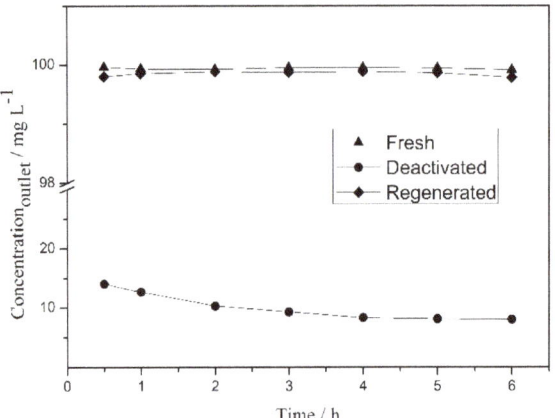

Figure 7. Dechlorination of 4-CP during HDC reaction in fresh, used and deactivated Pd/CNT-Ni foam micro reactors. Reaction conditions: residence time (3 min), Pd loading (0.5 wt.%), FA/substrate molar ratio (5:1), 4-chlorophenol concentration (1 mmol/L).

Lastly, a deactivated Pd/CNT-Ni foam micro reactor was regenerated for 3 h in an argon atmosphere (20 mL/min flow rate) at 250 °C. An almost complete recovery of the HDC catalytic activity was presented by the regenerated micro reactor (Figure 7), due to the efficient removal of the absorbed phenol layer from the Pd catalyst surface during the regeneration process (Figure 1f). This result demonstrated that the Pd/CNT-Ni foam micro reactor system can be regenerated efficiently through a simple heating process, facilitating its practical application significantly.

3. Materials and Methods

3.1. Preparation of Pd/CNT-Ni Foam Micro Reactor (Pd/CNT-Ni Foam Composite Catalyst)

A detailed preparation of the Pd/CNT-Ni foam micro reactor has been reported in the literature [35]. The active Pd was introduced into the micro channel of a monolithic CNT-Ni foam composite through the wetness incipient impregnation method. The pretreated CNT-Ni foam composite (about 15 wt.% CNTs) was impregnated with an acetone solution, then calcined at 300 °C for 2 h. Finally, the samples were reduced with 20 mL/min hydrogen flow at 300 °C for 2 h.

3.2. Catalytic Performance Evaluation

The HDC performance evaluations in micro reactors were carried out in a continuous flow Pd/CNT-Ni foam micro reactor system (Figure 8a) at room temperature and atmospheric pressure. Seven pieces of 65 × 6 × 1.8 mm (L × W × H) Pd/CNT-Ni foam micro reactor (about 1.5 g) were embedded into the stainless steel reaction chamber (Figure 8b) to assemble the reactor system with two fixed stainless steel plates. The 4-CP (1 mmol/L) and FA aqueous solution with an appropriate flow rate was made to flow into a micro reactor through a micro pump. The effluent liquid products of the reactor were collected for further analysis.

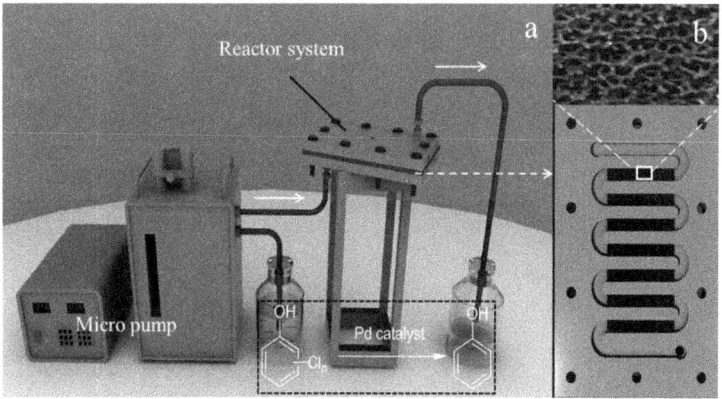

Figure 8. Sketched configuration of the continuous flow Pd/CNT-Ni foam micro reactor system (**a**) and the reaction chamber embedded with a Pd/CNT-Ni foam micro reactor (**b**).

We also evaluated the HDC performance in a continuous flow packed bed reactor with a Pd/CNT-Ni foam composite catalyst. Small pieces of Pd/CNT-Ni foam composite catalysts (3 × 3 × 1.8 mm) were packed into a 100-mm length of quartz tubing (8 mm I.D.) to assemble the packed bed reactor. The reagents aqueous solution was then pumped into the reactor, and the effluent liquid products of the reactor were collected for further analysis.

The effluent liquid samples were analyzed by high-performance liquid chromatography (HPLC) with a Zorbax XDB-C18 column (Agilent, USA) [1,12,35]. Inductively coupled plasma mass-spectrometry (ICP-MS) was employed to measure the Pd contents in the effluents.

3.3. Catalyst Characterization

A JSM-4800F electron microscope (JEOL Ltd., Tokyo, Japan) was employed for the field emission scanning electron microcopy (FESEM) analysis. To carry out transmission electron microscopy (TEM) and high resolution transmission electron microscopy (HRTEM) on A JEOL-2100F electron microscope (JEOL Ltd., Tokyo, Japan), the powder Pd/CNTs sample was prepared by separating it from the Ni foam skeleton through hydrochloric acid (0.1 M) immersion and ultrasonic treatment. The tests of Pd concentration in the effluents were conducted through Agilent 7500a (Agilent, USA) inductively coupled plasma mass-spectrometry.

4. Conclusions

The HDC of CPs using FA as a hydrogen source was investigated for the first time in a continuous flow micro reactor. A continuous flow Pd/CNT-Ni foam micro reactor system was developed for the rapid and highly efficient HDC of CPs. This micro reactor system, providing a micro size reaction channel (100–500 μm) and well-dispersed Pd nanoparticles (2.68 nm), exhibited a higher catalytic activity of HDC compared with the conventional packed bed reactor.

The influences of the Pd content and FA/substrate molar ratio on HDC performance in a continuous flow micro reactor were discussed. A very low FA/substrate molar ratio (5:1) and short reaction time (3 min) were required to obtain the desired outcomes (dichlorination >99.9%, 4-CP outlet concentration <0.1 mg/L) for a micro reactor with 0.5 wt.% Pd loading. Moreover, the long-term deactivation of HDC, which can be mainly attributed to the phenol accumulation on the Pd catalyst surface, was observed. The catalytic activity deactivation of this micro reactor system can be almost completely recovered by the efficient removal of the absorbed phenol layer from the Pd catalyst surface, making it a promising candidate for the HDC of wastewater containing highly toxic chlorinated organic pollutants and other Pd-catalyzed hydrogenation reactions.

This work shows that the usage of a micro reactor system is practicable and highly efficient for HDC in continuous flow. Future plans involve the development of a new Pd-based micro reactor system with designable structures and unique physical and chemical properties, which could remove target contaminants and enable the in situ regeneration of active components simultaneously.

Author Contributions: This study was conducted through contributions of all authors. J.X. designed the study, performed the experiments, and wrote the manuscript. Y.M. was involved in performing the experiments.

Funding: This work was supported by the Natural Science Foundation of China (grant number 21463030) and the Guizhou Province Technology Department of China (grant number QKHLHZ-2014-7550).

Conflicts of Interest: The authors declare no conflict of interest.

References

1. Lan, Y.; Yang, L.; Zhang, M.; Zhang, W.; Wang, S. Microreactor of Pd nanoparticles immobilized hollow microspheres for catalytic hydrodechlorination of chlorophenols in water. *Appl. Mater. Interfaces* **2010**, *2*, 127–133. [CrossRef] [PubMed]
2. Zhou, S.; Jin, X.; Sun, F.; Zhou, H.; Yang, C.; Xia, C. Combination of hydrodechlorination and biodegradation for the abatement of chlorophenols. *Water Sci. Technol.* **2012**, *65*, 780–786. [CrossRef] [PubMed]
3. Sun, C.; Wu, Z.; Mao, Y.; Yin, X.; Ma, L.; Wang, D.; Zhang, M. A highly active Pd on Ni–B bimetallic catalyst for liquid-phase hydrodechlorination of 4-chlorophenol under mild conditions. *Catal. Lett.* **2011**, *141*, 792–798. [CrossRef]
4. Cui, X.; Zuo, W.; Tian, M.; Dong, Z.; Ma, J. Highly efficient and recyclable Ni MOF-derived N-doped magnetic mesoporous carbon-supported palladium catalysts for the hydrodechlorination of chlorophenols. *J. Mol. Catal. A Chem.* **2016**, *423*, 386–392. [CrossRef]
5. Yin, L.F.; Shen, Z.Y.; Niu, J.F.; Jing, C.; Duan, Y.P. Degradation of pentachlorophenol and 2,4-dichlorophenol by sequential visible-light driven photocatalysis and laccase catalysis. *Environ. Sci. Technol.* **2010**, *44*, 9117–9122. [CrossRef] [PubMed]
6. Kao, C.M.; Chen, K.F.; Chen, Y.L.; Chen, T.Y.; Huang, W.Y. Biobarrier system for remediation of TCE-contaminated aquifers. *Bull. Environ. Contam. Toxicol.* **2004**, *72*, 87–93. [CrossRef] [PubMed]
7. Untea, I.; Orbeci, C.; Tudorache, E. Oxidative degradation of 4-chlorophenol from aqueous solution by photo-fenton advanced oxidation process. *Environ. Eng. Manag. J.* **2006**, *5*, 661–674. [CrossRef]
8. Suárez-Ojeda, M.E.; Fabregat, A.; Stüber, F.; Fortuny, A.; Carrera, J.; Font, J. Catalytic wet air oxidation of substituted phenols: Temperature and pressure effect on the pollutant removal, the catalyst preservation and the biodegradability enhancement. *Chem. Eng. J.* **2007**, *132*, 105–115. [CrossRef]
9. Ma, X.; Liu, Y.; Li, X.; Xu, J.; Gu, G.; Xia, C. Water: The most effective solvent for liquid-phase hydrodechlorination of chlorophenols over Raney Ni catalyst. *Appl. Catal. B Environ.* **2015**, *165*, 351–359. [CrossRef]
10. Zhang, W.; Wang, F.; Li, X.; Liu, Y.; Ma, J. Pd nanoparticles modified rod-like nitrogen-doped ordered mesoporous carbons for effective catalytic hydrodechlorination of chlorophenols. *RSC Adv.* **2016**, *6*, 27313–27319. [CrossRef]
11. Xia, C.; Liu, Y.; Zhou, S.; Yang, C.; Liu, S.; Xu, J.; Yu, J.; Chen, J.; Liang, X. The Pd-catalyzed hydrodechlorination of chlorophenols in aqueous solutions under mild conditions: A promising approach to practical use in wastewater. *J. Hazard. Mater.* **2009**, *169*, 1029–1033. [CrossRef] [PubMed]
12. Lan, L.; Du, F.; Xia, C. The reaction mechanism for highly effective hydrodechlorination of p-chlorophenol over a Pd/CNTs catalyst. *RSC Adv.* **2016**, *6*, 109023–109029. [CrossRef]
13. Fang, X.; Fang, D. Performance of palladium–tin bimetallic catalysts supported on activated carbon for the hydrodechlorination of 4-chlorophenol. *RSC Adv.* **2017**, *7*, 40437–40443. [CrossRef]

14. Yu, X.; Wu, T.; Yang, X.J.; Xu, J.; Auzam, J.; Semiat, R.; Han, Y.F. Degradation of trichloroethylene by hydrodechlorination using formic acid as hydrogen source over supported Pd catalysts. *J. Hazard. Mater.* **2016**, *305*, 178–189. [CrossRef] [PubMed]
15. Kopinke, F.D.; Mackenzie, K.; Köhler, R. Catalytic hydrodechlorination of groundwater contaminants in water and in the gas phase using Pd/γ-Al$_2$O$_3$. *Appl. Catal. B Environ.* **2003**, *44*, 15–24. [CrossRef]
16. Gómez-Quero, S.; Cárdenas-Lizana, F.; Keane, M.A. Solvent effects in the hydrodechlorination of 2,4-dichlorophenol over Pd/Al$_2$O$_3$. *AIChE J.* **2010**, *56*, 756–767. [CrossRef]
17. Gómez-Quero, S.; Cárdenas-Lizana, F.; Keane, M.A. Effect of metal dispersion on the liquid-phase hydrodechlorination of 2,4-dichlorophenol over Pd/Al$_2$O$_3$. *Ind. Eng. Chem. Res.* **2008**, *47*, 6841–6853. [CrossRef]
18. Chang, W.; Kim, H.; Lee, G.Y.; Ahn, B.J. Catalytic hydrodechlorination reaction of chlorophenols by Pd nanoparticles supported on grapheme. *Res. Chem. Intermed.* **2016**, *42*, 71–82. [CrossRef]
19. Díaz, E.; McCall, A.; Faba, L.; Sastre, H.; Ordóñez, S. Trichloroethylene hydrodechlorination in water using formic acid as hydrogen source: Selection of catalyst and operation conditions. *Environ. Prog. Sustain.* **2013**, *32*, 1217–1222. [CrossRef]
20. Kopinke, F.D.; Mackenzie, K.; Koehler, R.; Georgi, A. Alternative sources of hydrogen for hydrodechlorination of chlorinated organic compounds in water on Pd catalysts. *Appl. Catal. A Gen.* **2004**, *271*, 119–128. [CrossRef]
21. State, R.; Papa, F.; Tabakova, T.; Atkinson, I.; Negrila, C.; Balint, I. Photocatalytic abatement of trichlorethylene over Au and Pd–Au supported on TiO$_2$ by combined photomineralization/hydrodechlorination reactions under simulated solar irradiation. *J. Catal.* **2017**, *346*, 101–108. [CrossRef]
22. Di Sarli, V.; Di Benedetto, A. Effects of non-equidiffusion on unsteady propagation of hydrogen-enriched methane/air premixed flames. *Int. J. Hydrog. Energy* **2013**, *38*, 7510–7518. [CrossRef]
23. Calvo, L.; Gilarranz, M.A.; Casas, J.A.; Mohedano, A.F.; Rodriguez, J.J. Hydrodechlorination of 4-chlorophenol in water with formic acid using a Pd/activated carbon catalyst. *J. Hazard. Mater.* **2009**, *161*, 842–847. [CrossRef]
24. Cellier, P.P.; Spindler, J.F.; Taillefer, M.; Cristau, H.J. Pd/C-catalyzed room-temperature hydrodehalogenation of aryl halides with hydrazine hydrochloride. *Tetrahedron Lett.* **2003**, *44*, 7191–7195. [CrossRef]
25. Xia, C.; Liu, Y.; Zhou, S.; Yang, C.; Liu, S.; Guo, S.; Liu, Q.; Yu, J.; Chen, J. The influence of ion effects on the Pd-catalyzed hydrodechlorination of 4-chlorophenol in aqueous solutions. *Catal. Commun.* **2009**, *10*, 1443–1445. [CrossRef]
26. Dong, Z.; Le, X.; Liu, Y.; Dong, C.; Ma, J. Metal organic framework derived magnetic porous carbon composite supported gold and palladium nanoparticles as highly efficient and recyclable catalysts for reduction of 4-nitrophenol and hydrodechlorination of 4-chlorophenol. *J. Mater. Chem. A* **2014**, *2*, 18775–18785. [CrossRef]
27. Shao, Y.; Xu, Z.; Wan, H.; Wan, Y.; Chen, H.; Zheng, S.; Zhu, D. Enhanced liquid phase catalytic hydrodechlorination of 2,4-dichlorophenol over mesoporous carbon supported Pd catalysts. *Catal. Commun.* **2011**, *12*, 1405–1409. [CrossRef]
28. Molina, C.B.; Pizarro, A.H.; Casas, J.A.; Rodriguez, J.J. Aqueous-phase hydrodechlorination of chlorophenols with pillared clays-supported Pt, Pd and Rh catalysts. *Appl. Catal. B Environ.* **2014**, *148–149*, 330–338. [CrossRef]
29. Hartman, R.L.; McMullen, J.P.; Jensen, K.F. Deciding whether to go with the flow: Evaluating the merits of flow reactors for synthesis. *Angew. Chem. Int. Ed.* **2011**, *50*, 7502–7519. [CrossRef]
30. Gutmann, B.; Cantillo, D.; Kappe, C.O. Continuous-flow technology-a tool for the safe manufacturing of active pharmaceutical ingredients. *Angew. Chem. Int. Ed.* **2015**, *54*, 6688–6728. [CrossRef]
31. Naber, J.R.; Buchwald, S.L. Packed-bed reactors for continuous-flow C-N cross-coupling. *Angew. Chem. Int. Ed.* **2010**, *49*, 9469–9474. [CrossRef] [PubMed]
32. Kumar, U.; Panda, D.; Biswas, K.G. Non-lithographic copper-wire based fabrication of micro-fluidic reactors for biphasic flow applications. *Chem. Eng. J.* **2018**, *344*, 221–227. [CrossRef]
33. Tsubogo, T.; Ishiwata, T.; Kobayashi, S. Asymmetric carbon-carbon bond formation under continuous-flow conditions with chiral heterogeneous catalysts. *Angew. Chem. Int. Ed.* **2013**, *52*, 6590–6604. [CrossRef] [PubMed]
34. Rossetti, I. Continuous flow (micro-)reactors for heterogeneously catalyzed reactions: Main design and modelling issues. *Catal. Today* **2018**, *308*, 20–31. [CrossRef]

35. Xiong, J.; Ma, Y.; Yang, W.; Zhong, L. Rapid, highly efficient and stable catalytic hydrodechlorination of chlorophenols over novel Pd/CNTs-Ni foam composite catalyst in continuous-flow. *J. Hazard. Mater.* **2018**, *355*, 89–95. [CrossRef] [PubMed]
36. Liu, W.; Rodriguez, P.; Borchardt, L.; Foelske, A.; Yuan, J.; Herrmann, A.K.; Geiger, D.; Zheng, Z.; Kaskel, S.; Gaponik, N.; et al. Bimetallic aerogels: High-performance electrocatalysts for the oxygen reduction reaction. *Angew. Chem. Int. Ed.* **2013**, *52*, 9849–9852. [CrossRef] [PubMed]
37. Ding, H.; Shi, X.Z.; Shen, C.M.; Hui, C.; Xu, Z.C.; Li, C.; Tian, Y.; Wang, D.K.; Gao, H.J. Synthesis of monodisperse palladium nanocubes and their catalytic activity for methanol electrooxidation. *Chin. Phys. B* **2010**, *19*, 106104.
38. Jadbabaei, N.; Ye, T.; Shuai, D.; Zhang, H. Development of palladium-resin composites for catalytic hydrodechlorination of 4-chlorophenol. *Appl. Catal. B Environ.* **2017**, *205*, 576–586. [CrossRef]
39. Casanovas, A.; Domínguez, M.; Ledesma, C.; López, E.; Llorca, J. Catalytic walls and micro-devices for generating hydrogen by low temperature steam reforming of ethanol. *Catal. Today* **2009**, *143*, 32–37. [CrossRef]
40. Xiong, J.; Dong, X.; Dong, Y.; Hao, X.; Hampshire, S. Dual-production of nickel foam supported carbon nanotubes and hydrogen by methane catalytic decomposition. *Int. J. Hydrog. Energy* **2012**, *37*, 12307–12316. [CrossRef]
41. Yu, Z.; Chen, D.; Tøtdal, B.; Zhao, T.; Dai, Y.; Yuan, W.; Holmen, A. Catalytic engineering of carbon nanotube production. *Appl. Catal. A. Gen.* **2005**, *279*, 223–233. [CrossRef]
42. Yuan, G.; Keane, M.A. Liquid phase hydrodechlorination of chlorophenols over Pd/C and Pd/Al$_2$O$_3$: A consideration of HCl/catalyst interactions and solution pH effects. *Appl. Catal. B Environ.* **2004**, *52*, 301–314. [CrossRef]
43. Keane, M.A. A review of catalytic approaches to waste minimization: Case study-liquid-phase catalytic treatment of chlorophenols. *J. Chem. Technol. Biotechnol.* **2005**, *80*, 1211–1222. [CrossRef]
44. Gómez-Quero, S.; Cárdenas-Lizana, F.; Keane, M.A. Liquid phase catalytic hydrodechlorination of 2,4-dichlorophenol over Pd/Al$_2$O$_3$: Batch vs. continuous operation. *Chem. Eng. J.* **2011**, *166*, 1044–1051. [CrossRef]
45. Singh, P.; Kulkarni, M.V.; Gokhale, S.P. Enhancing the hydrogen storage capacity of Pd-functionalized multi-walled carbon nanotubes. *Appl. Surf. Sci.* **2012**, *258*, 3405–3409. [CrossRef]

© 2019 by the authors. Licensee MDPI, Basel, Switzerland. This article is an open access article distributed under the terms and conditions of the Creative Commons Attribution (CC BY) license (http://creativecommons.org/licenses/by/4.0/).

Article

Mg-Catalyzed OPPenauer Oxidation—Application to the Flow Synthesis of a Natural Pheromone

Virginie Liautard, Mélodie Birepinte, Camille Bettoli and Mathieu Pucheault *

Institut des Sciences Moléculaires (ISM), UMR 5255 CNRS—Université de Bordeaux, 351 Cours de la Libération, 33405 Talence CEDEX, France; virginie.liautard@u-bordeaux.fr (V.L.); melodie.birepinte@u-bordeaux.fr (M.B.); camille.bettoli@etu.u-bordeaux.fr (C.B.)
* Correspondence: mathieu.pucheault@u-bordeaux.fr

Received: 3 October 2018; Accepted: 2 November 2018; Published: 8 November 2018

Abstract: The so-called OPPenauer oxidation is well known for its ability to oxidize valuable alcohols into their corresponding aldehydes or ketones. In particular, it has proven to be extremely successful in the oxidation of sterols. On the other hand, its application—in the original formulation—to the obtainment of ketones outside the field of steroids met a more limited success because of less favorable thermodynamics and side reactions. To circumvent these issues, the first example of magnesium-catalyzed OPPenauer oxidation is described. The oxidation of primary and secondary alcohol was performed using pivaldehyde or bromaldehyde as the oxidant and cheap magnesium *tert*-butoxide as catalyst. Decent to excellent yields were obtained using reasonable catalytic charge. The synthesis of a pheromone stemming from the *Rhynchophorus ferrugineus* was obtained by tandem addition-oxidation of 2-methylpentanal and the process was successfully applied to continuous flow on a multigram scale.

Keywords: Oppenauer oxidation; magnesium; catalysis; alcohols; aldehydes; ketones; pheromone; *Rhynchophorus ferrugineus*

---

1. Introduction

OPPenauer (OPP) oxidation [1], the reverse process of the Meerwein-Ponndorf-Verley (MPV) reduction [2,3] is a classical, well known yet useful method. Indeed, alcohol oxidation to carbonyl constitutes one of the most important transformations in organic chemistry. However, despite the first initial success of the OPP oxidation procedure in the oxidation of steroidal compounds, it did not find widespread utility in the organic chemistry field outside of a few tuned and very specific natural product syntheses [4–9]. This is mostly because over-stoichiometric amounts of aluminum reagents were required to achieve good yields under reasonable reaction conditions [10–15]. To overcome this sluggish activity, some typical procedures involving a broad range of organic oxidizing reagents and metal-based systems have emerged. Simple salts as K [16,17] and Na [18,19] were first used in respectively stoichiometric and catalytic quantities to promote the OPP oxidation while more complex metal-based systems rapidly arose. Most of them were involving transition metals such as Ru [20,21], Ir [22,23], Fe [24], Zr [25] or more recently Mn [26], used in catalytic amounts but also other elements such as In [27,28], or Si [29]. These advances have contributed to the re-emergence of the OPP oxidation as an attractive process. Nevertheless, all these procedures have disadvantages; they rely on expensive metal complexes and/or elevated temperatures, are often toxic and eventually generate noxious wastes. Among these metals, magnesium appears to be an extremely attractive candidate as it possesses many properties appealing in catalysis such as low cost, high abundance, large possibilities of coordination and easy treatment of the resulting salts.

The Mg-OPP oxidation was first explored as a stoichiometric process in 1987 by Brian Byrne et al. using excess of magnesium under the form of Grignards [30], followed by numerous

Knochel's protocols which were considered to be the most advantageous as the *in situ* formed magnesium alkoxides are intrinsically powerful promoters for the forthcoming hydride transfer [31–33]. Nevertheless, Grignard reagents should be accurately titrated beforehand and transferred carefully in case over- or less-loading will lead to significant side reactions such as Aldol condensation and Tishchenko esterifications [34]. Additionally, routes starting from Grignard reagents demand low reaction temperatures, and sensitive functional groups are still found difficult to tolerate [35]. To circumvent these problems, we believe that a magnesium catalyzed variant of the well-known OPP oxidation could be a good way to promote a more effective system to oxidize alcohols into carbonyls without the disagreement of the other systems.

2. Results and Discussions

The oxidation of ferrugineol (**1**) (see Figure S1), to give the corresponding ketone ferrugineone (**2**) was chosen to be our model substrate (see Figure S2). Indeed, this secondary alcohol is challenging since it is sterically hindered, non-conjugated, and not easily enolizable; however, it can lead to elimination reactions in basic medium which would give the corresponding alkene.

2.1. Preliminary Tests

Pivaldehyde [14] was selected as the first hydride acceptor owing to its low dielectric constant and high oxidation potential (E_0 = 211 mV) [19]. Different magnesium oxides and a magnesium salt were studied to see if the OPP oxidation could be performed with catalytic amounts of these species (Table 1).

Table 1. Screening of conditions for the Mg-catalyzed oxidation of ferrugineol (**1**) to ferrugineone (**2**).

Entry	"Mg" (equiv)	Temperature (°C)	Time (h)	^{1}H NMR (%) Conversion
entry 1	MgCl$_2$ (0.4)	100	3	3
entry 2	Mg(OH)$_2$ (0.4)	100	3	2
entry 3	Mg(OEt)$_2$ (0.4)	100	3	31
entry 4	Mg(OEt)$_2$ (0.4)	100	6	33
entry 5	MgO (0.4)	100	3	2
entry 6	**Mg(O*t*Bu)$_2$ (0.3)**	**110**	**1**	**70**
entry 7	Mg(O*t*Bu)$_2$ (0.3)	60	1	21
entry 8	Mg(O*t*Bu)$_2$ (0.3)	60	3	32

Reaction conditions: ferrugineol (1.0 equiv; 0.63 mmol), pivaldehyde (2.0 equiv), Mg catalyst (0.3 or 0.4 equiv), C$_6$D$_6$ (400 µL) under nitrogen. Note: Every bold used in tables highlight the best result.

Different catalytic charges were tested in deuterated benzene to evaluate the reaction efficiency via NMR conversions. It clearly appeared that MgCl$_2$, Mg(OH)$_2$ and MgO as catalysts were not the best candidates given that 3% conversion was the maximum obtained after 3 h heating (Table 1, entries 1, 2 and 5, Figures S3, S4 and S7).

Mg(OEt)$_2$ gave average conversions (Table 1, entries 3 and 4, Figures S5 and S6) whereas magnesium *tert*-butoxide appeared to be the most effective to catalyze the reaction (entry 6, Figures S8 and S9). It is noteworthy that the resulting *t*BuOH cannot be oxidized therefore limiting side reactions (Figures S10, S11, S12 and S13). Reasonable NMR conversion was observed (70%) after only an hour 30% catalyst with a slight aldehyde excess at 110 °C. It seems likely that the reaction proceeds through deprotonation, alcoholate-adduct formation via hydride transfer followed by classical OPP oxidation (Scheme 1).

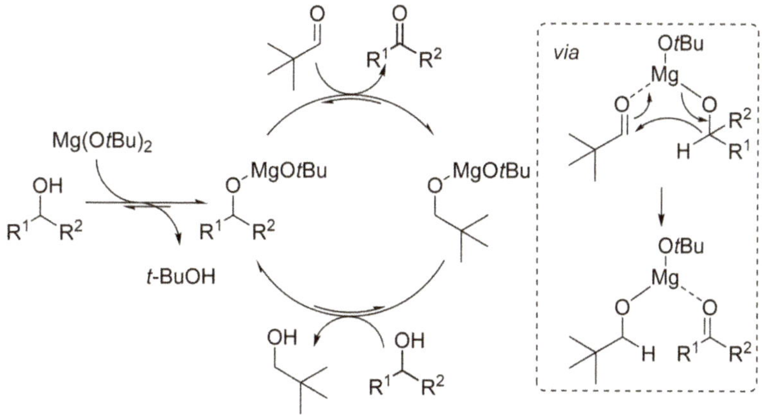

Scheme 1. Proposed mechanism for the Mg-catalyzed OPP oxidation.

2.2. Optimization of the Reaction Conditions

2.2.1. Aldehyde and Temperature Studies

Magnesium *tert*-butoxide was thus chosen as the magnesium source whereas the aldehyde, the solvent and temperature, and the alcohol nature were studied. Pivaldehyde was first chosen as the hydride acceptor as described earlier but the influence of the aldehyde's nature was also studied (Table 2, entries 1 to 5). 2-methylpentanal and isobutyraldehyde are inexpensive, enolizable, yet sterically hindered aldehydes preventing from side reaction. They both led to low conversions, respectively 21 and 25% (Table 2, entries 2 and 3). Pivaldehyde and bromaldehyde were more efficient than most of the other aldehydes (Table 2, entries 1 and 5) with respectively 70 and 84% conversions after only one hour heating so they were both further studied.

Table 2. Aldehydes tests results.

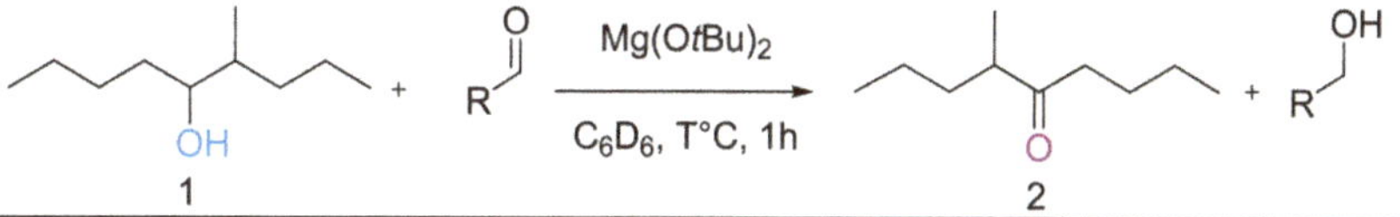

Entry	Aldehyde	Temperature (°C)	^{1}H NMR (%) Conversions [1]
entry 1	Pivaldehyde	110	70
entry 2	2-methylpentanal	60	21
entry 3	isobutyraldehyde	100	25
entry 4	Cyclohexane carboxaldehyde	100	54
entry 5	**Bromaldehyde**	100	**84**

Reaction conditions: ferrugineol (1.0 equiv; 0.63 mmol), aldehyde (2.0 equiv), Mg (0.3 equiv), C$_6$D$_6$ (400 µL) under nitrogen. [1] Conversions are calculated according to the starting material consumption related to the appearance of product. Note: Every bold used in tables highlight the best result.

2.2.2. Solvent Optimization

The optimal conditions of solvent and temperature were found by testing the oxidation of both ferrugineol and cyclohexanol as they are illustrating different types of substrate (Table 3). Unsurprisingly, considering the preliminary results in C_6D_6, non-polar, non-protic solvents such as toluene led to the best results (up to 71%). Interestingly, undistilled solvent were giving higher conversions than the distilled ones (Table 3, entries 4, 7 and 8), meaning that the presence of water could have an influence on the reaction efficiency. Known quantities of water (0.5 to 2.0 equiv) were then added to dry C_6D_6 to study this influence but it actually conducted to lower conversions (Table 3, entries 12 to 14). Further measurement of water trace amounts in both non-distilled C_6D_6 and toluene using a Karl Fisher metrohm were also performed. Indeed, adding 0.5 to 2.0 equiv of water actually meant 5.7 to 22 mg while we found that there were only 1.1 ppm (4.9×10^{-4} equiv) and 1.9 ppm (8.4×10^{-4} equiv) of water in 400 µL of respectively C_6D_6 and toluene. These values tell that traces amount of water are enough to increase the reaction conversion (probably involving the establishment of hydrogen bonds) whereas too much water would actually be unfavorable. Knowing that, the optimized reaction conditions were fixed with 30 mol% Mg(OtBu)$_2$, 2.0 equiv of pivaldehyde in C_6D_6 or toluene, according to the scale, for 1 h at 80 °C.

Table 3. Solvent optimization results.

Entry	Alcohol	Aldehyde	Solvent	^{1}H NMR (%) Conversions ^{1}
entry 1	Ferrugineol	Pivaldehyde	C_6D_6 [a]	70
entry 2	Ferrugineol	Pivaldehyde	Toluene [b]	54
entry 3	Ferrugineol	Pivaldehyde	DMF [b]	16
entry 4	Ferrugineol	Pivaldehyde	THF [b]	3
entry 5	Ferrugineol	Pivaldehyde	MeTHF [b]	16
entry 6	Ferrugineol	Pivaldehyde	DME [b]	traces
entry 7	Cyclohexanol	Pivaldehyde	C_6D_6 [b]	62
entry 8	**Cyclohexanol**	**Pivaldehyde**	**Toluene [c]**	**71**
entry 9	Cyclohexanol	Pivaldehyde	Dry Toluene [c]	46
entry 10	Cyclohexanol	Pivaldehyde	DCE [c]	55
entry 11	Cyclohexanol	Pivaldehyde	MTBE [c]	57
entry 12	Cyclohexanol	Bromaldehyde	C_6D_6 [d]	60
entry 13	Cyclohexanol	Bromaldehyde	Dry C_6D_6 [d]	31
entry 14	Cyclohexanol	Bromaldehyde	Dry C_6D_6 + H_2O [d]	traces
entry 15	Cyclohexanol	Bromaldehyde	Toluene [c]	45
entry 16	Cyclohexanol	Bromaldehyde	DCE [c]	36
entry 17	Cyclohexanol	Bromaldehyde	MTBE [c]	38
entry 18	Cyclohexanol	Bromaldehyde	Dry THF [c]	3
entry 19	Cyclohexanol	Bromaldehyde	Dry DCM [d]	39
entry 10	Cyclohexanol	Bromaldehyde	Dry Toluene [d]	49

Reaction conditions: alcohol (1.0 equiv), aldehyde (2.0 equiv), Mg catalyst (0.3 equiv), heating 1 h under nitrogen. [a] 110 °C, [b] 100 °C, [c] 80 °C, [d] 60 °C. ^{1} Conversions are calculated according to the alcohol consumption.

To demonstrate the generality and scope of this protocol we applied our conditions to different types of alcohols for the preparation of aromatic, aliphatic and α-β-saturated ketones (Figure 1).

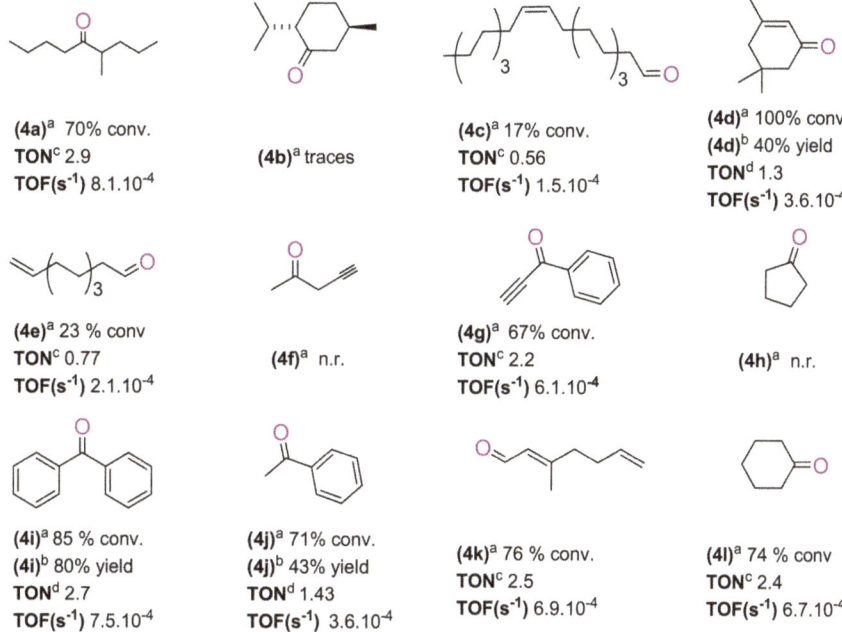

Figure 1. Results for the Mg-catalyzed OPP oxidation of various ketones. General Reaction conditions: Alcohol (**3a**) to (**3l**) (1.0 equiv), Mg(O*t*Bu)$_2$ (0.3 equiv), aldehyde (2.0 equiv) [a] Procedure A; results reported as NMR conversions based on the starting alcohol. [b] Procedure B; results reported as isolated yields after column chromatography on silica gel. [c] TON (turn over number) and TOF (turn over frequency) were calculated using conversion values. [d] TON and TOF were calculated using isolated yields values.

2.3. Batch Applications

Alcohols were oxidized into their corresponding ketones with decent to high conversions (up to 100%, with a TON of 3.3). When the *ortho* position was too hindered (**4b**) or involved in a strained cycle (**4h**) the product was only observed as traces. Reaction was also limited by poor solubility (**4c**) or high volatility of the product, sometimes lost during isolation (**4f**). Subsequent transformation of the resulting products as hydrazones for low boiling points products was attempted but did not helped for any isolation. A direct distillation from the reaction mixture could be envisioned as an alternative. **4d** and **4i** were isolated after purification via column chromatography on silica gel.

Overall, despite some limitations, this Mg-catalyzed method proved to be efficient on different substrates.

2.4. Flow Application

2.4.1. Using *n*BuMgCl in Catalytic Amounts

One of the objectives of developing this procedure was to synthesize ferrugineone starting from the corresponding alcohol ferrugineol. These natural pheromones from the Red Palm Weevil (*Rhynchophorus ferrugineus*) are both used in biocontrol [36,37]. The sexual confusion applications require a 90/10 mixture of ferrugineol and ferrugineone [38]. We envisioned preparing such a mixture by tandem addition—oxidation using 2-methylpentanal and *n*BuMgCl. Using our oxidation conditions, the resulting conversion of ferrugineol, close to 70%, is satisfying. However, in the tandem process the system would be slightly different with alkylmagnesium alkoxide being produced in lieu of the dialkoxide. Therefore, a quick survey of nBuMgCl as catalyst was performed, rapidly screening different solvents. It promptly appeared that a catalytic version of the reaction was not possible as the best results obtained were around 20% conversion using 0.4 equiv of nBuMgCl (Table 4, entry 5) and increasing this amount to 0.5 equivalents did not improve the conversion greatly. The best solvent appeared to be C_6D_6 (Table 4, entry 5) but running the reaction neat gave similar conversion (Table 4, entry 2).

Table 4. nBuMgCl used as catalyst of the ferrugineol/ferrugineone oxidation.

Entry	nBuMgCl (equiv)	Solvent	^{1}H NMR (%) Conversions [1]
entry 1	0.1 to 0.3	neat	NR
entry 2	0.4	neat	18
entry 3	0.5	neat	traces
entry 4	0.4	THF	NR
entry 5	0.4	C_6D_6	21
entry 6	0.4	$CDCl_3$	traces
entry 7	0.4	DME	traces
entry 8	0.4	DMF	traces

Reaction conditions: ferrugineol (1.0 equiv), aldehyde (3.0 equiv), nBuMgCl (0.1 to 0.5 equiv), heating 1 h at 60 °C under inert atmosphere. [1] Conversions are calculated according to the starting material consumption related to the appearance of product.

2.4.2. Proposed Synthesis of the 90/10 Mixture

Flow chemistry is allowing for a better control of the reaction conditions, thus avoiding byproducts resulting from elimination or crossed aldol reactions. It also prevents thermal runaway leading to undesired side reactions. In our systems, we would easily separate the addition of nBuMgCl on the aldehyde, and the subsequent oxidation of the resulting ferrugineol using the excess of aldehyde injected in the system. In that case, 2-methylpentanal would serve as both the substrate and the oxidant (Scheme 2).

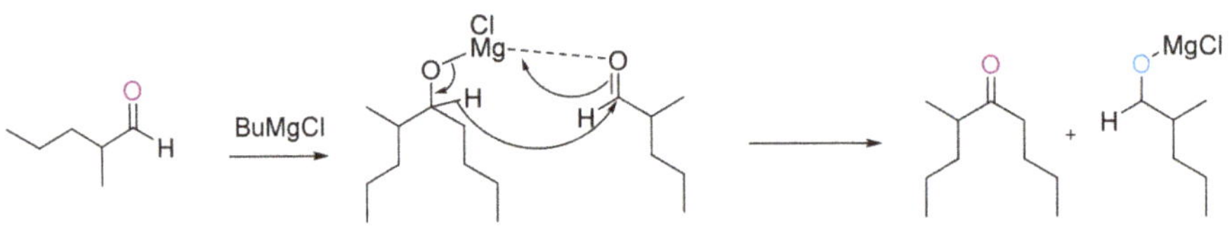

Scheme 2. Tandem process of the *in situ* Oppenauer oxidation of ferrugineol.

Applying this synthesis in flow chemistry allowed us to quickly evaluate the correlation of aldehyde excess and the alcohol/ketone ratio. However, the number of equivalents of nBuMgCl still needed to be optimized, as well as the reaction conditions for flow application.

2.4.3. Optimization of Flow Conditions

A series of experiments using various temperatures or various amounts of nBuMgCl was performed on an E-Series of Vapourtec. As decreasing the temperature below 60 °C led to lower conversions while increasing the temperature to 70 or 80 °C did not show any evidence of improvement, we decided to process with 60 °C for the rest of the optimization.

Different amounts of *n*BuMgCl (from 0.70 to 1.10 equiv.) were tested in flow chemistry to demonstrate which conditions would be the best to get to the expected 90/10 mixture of ferrugineol/ferrugineone. The experiments were followed in GC/MS using a DB-1701 capillary column and a HP 5973 mass selective detector (EI).

Decreasing the amount from 1.10 equiv to 0.90 equiv of *n*BuMgCl led to a mixture closed to the right proportions of ferrugineol/ferrugineone (Table 5, entry 3). Finally, we found that 2-methylpentanal (1.0 equiv), *n*BuMgCl (0.90 equiv), in the absence of solvent with a 4.54 mL tubing (3 min residence time) heated to 60 °C led to the formation of the ferrugineol/ferrugineone as an 86/14 mixture in 91% yield. Over the course of 8 h of operation and with a 3 min residence time, using relatively low flow rates, 173 g of the ferrugineol/ferrugineone mixture were obtained (Figure 2).

Table 5. Optimization of flow synthesis conditions.

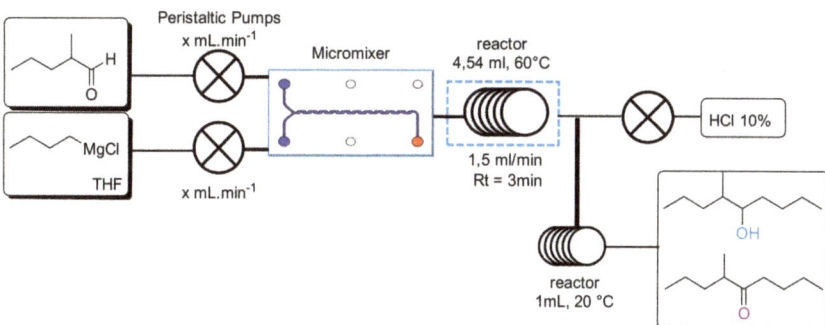

Entry	Aldehyde Flow (mL/min)	nBuMgCl Flow (mL/min)	nBuMgCl Conc.	nBuMgCl	% (2) (GC/MS)	% (1) (GC/MS)	Yield (%)
entry 1	0.281	1.22	2.04	1.10	0.7	99.3	60
entry 2	0.297	1.2	2.04	1.03	1.8	98.2	79
entry 3	0.329	1.17	2.04	0.90	14.0	86.0	91
entry 4	0.4	1.11	2.04	0.70	21.0	78.4	97

Reaction conditions: ferrugineol (1.0 equiv), aldehyde (3.0 equiv), *n*BuMgCl (0.70 to 1.10 equiv.), heating 8 h 30 at 60 °C with a residence time of 3 min. (**1**) is for ferrugineol and (**2**) is for ferrugineone.

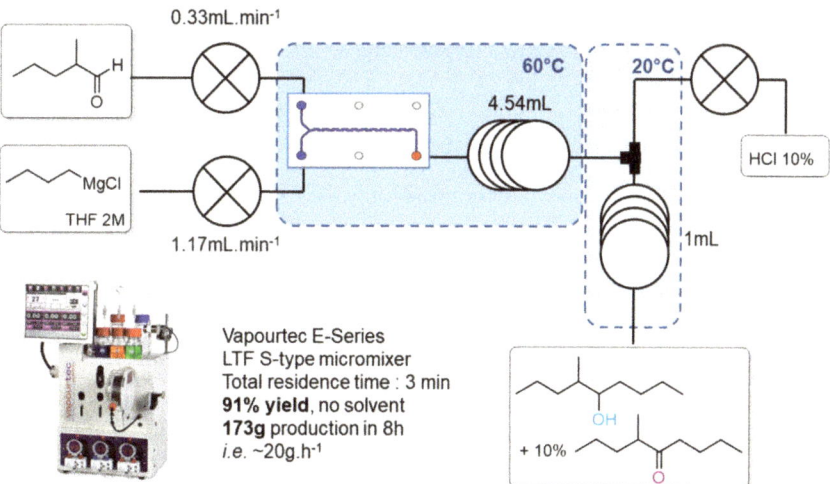

Figure 2. Flow chemistry process for the synthesis of a 90/10 ferrugineol/ferrugineone mixture.

3. Materials and Methods

All reagents and solvents were purchased from Sigma-Aldrich, (Sigma-Aldrich Chimie SARL, Tharabie France). Tribromoacetaldehyde was purchased from TCI (TCI Europe N.V., Zwijndrecht, Belgium). THF, dichloromethane, toluene and C_6D_6 were dried over sodium/benzophenone and freshly distilled under an atmosphere of argon before use. All commercially available solvents and reagents were use directly as received unless specified. All the laboratory glassware was dried in oven and cooled under vacuum before use.

Analytical thin layer chromatography (TLC) was carried out using 0.25 mm silica plates purchased from Merck. Eluted plates were visualized using aqueous $KMnO_4$ ($KMnO_4$ 3 g, K_2CO_3 20 g, aqueous 5% NaOH 5 mL, H_2O 300 mL). Silica gel chromatography was performed using 230–400 mesh silica gel purchased from Merck (Merck, Darmstadt, Germany) and more precisely from the supplier Sigma Aldrich (Sigma-Aldrich Chimie SARL, Tharabie France).

GC-MS analysis was performed with a HP 6890 series GC-system equipped with a J&W Scientific DB-1701 capillary column from Agilent (Agilent Technologies France, Les Ulis, France) and a HP 5973 mass selective detector (EI) also from Agilent (see above) using the following method: the temperature was held at 70 °C for 1 min, then the temperature increased till 230 °C with a heating rate of 20 °C/min and kept for 6 min at 230 °C.

^1H NMR was recorded on a 300 MHz, 400 MHz, and 600 MHz using Bruker Advance 300, Advance 400, and Advance 600, respectively (Bruker France S.A.S., Palaiseau, France). Chemical shifts (δ) are given in ppm relative to tetramethylsilane (external standard). ^{13}C NMR was recorded on a 300 MHz, 400 MHz, and 600 MHz using Brucker Advance 300, Advance 400, and Advance 600, respectively. Chemical shifts (δ) are given in ppm relative to tetramethylsilane (external standard).

Water traces amounts were measured using a TitroLine® 7500 KF trace from Thermo Fisher Scientific (Fisher Scientific SAS, Illkirch, France) with three injections using 400µL of the solvent studied each time and the average value was reported.

A Vapourtec E-series (V-3 perilstatic pumps) flow reactor was used for flow chemistry experiments (Vapourtec Ltd., Bury St Edmunds, Suffolk, UK). The reactor consists of a 4.54 mL 1/16″ PTFE tubing (0.81 mm I.D.) heated at 60 °C followed by a 1 mL 1/16″ PTFE tubing (0.81 mm I.D.). The E-Series come with a touchscreen interface, mounted at an ergonomically optimal height with full tilt adjustment. It allows to easily set the key flow rates but also the temperature parameters which can be accurately controlled ($\pm$ 1 °C), through a feedback system, in the range of room temperature −150 °C. (https://www.vapourtec.com/products/e-series-flow-chemistry-system-overview/).

In a typical calculation, the NMR conversion of alcohol in ketone was evaluated using the following equation and the error of measurement of the NMR is supposed to be ±5%:

$$\text{Conversion \%} = \frac{\frac{I\text{ketone}}{\text{number of protons considered}}}{\frac{I\text{ketone} + I\text{alcohol}}{\text{number of protons considered}}} \times 100$$

where I is the integral value of protons from alcohol and ketone in the spectrum of reaction mixture.

4. Conclusions

Overall, we developed an efficient and selective method for the oxidation of various alcohols to the corresponding aldehydes and ketones under mild conditions. This unprecedented Mg-catalyzed OPP oxidation highlights the efficiency and importance of the OPP method for the oxidation of primary and secondary alcohols and in the same time illustrates interesting versatile reactivity of magnesium derivatives. This protocol has notable advantages as it mostly uses common and inexpensive chemicals, operates with short reaction times especially using flow systems, gives good yields, lack of usual byproducts and can be applied to the oxidation of many alcohols. In addition, studying the reactivity of Grignard's reagents associated with the main advantages of continuous flow chemistry led to the development of a new procedure to selectively synthesize more than 170 g of a natural pheromone mixture with a perfect control of the ratio owing to the flexibility of the system.

Supplementary Materials: The following are available online at http://www.mdpi.com/2073-4344/8/11/529/s1, Figure S1: ^{1}H NMR of pure Ferrugineol (**1**), Figure S2: ^{1}H NMR of pure Ferrugineone (**2**), Figure S3: ^{1}H NMR of Table 1 Entry 1. MgCl$_2$ (0.4 equiv), 100 °C, 3 h, Figure S4: ^{1}H NMR of Table 1 Entry 2. Mg(OH)$_2$ (0.4 equiv), 100 °C, 3 h, Figure S5: ^{1}H NMR of Table 1 Entry 3. Mg(OEt)$_2$ (0.4 equiv), 100 °C, 3 h, Figure S6: ^{1}H NMR of Table 1 Entry 4. Mg(OEt)$_2$ (0.4 equiv), 100 °C, 6 h, Figure S7: ^{1}H NMR of Table 1 Entry 5. MgO (0.4 equiv), 100 °C, 3 h, Figure S8: ^{1}H NMR of Table 1 Entry 6. Mg(OtBu)$_2$ (0.3 equiv), 110 °C, 1 h, Figure S9: ^{1}H NMR of Table 1 Entry 6—Zoom from 5ppm to 0ppm—Protons attribution, Figure S10: ^{1}H NMR of Table 1 Entry 7. Mg(OtBu)$_2$ (0.3 equiv), 60 °C, 1 h, Figure S11: ^{1}H NMR of Table 1 Entry 8. Mg(OtBu)$_2$ (0.3 equiv), 60 °C, 3 h, Figure S12: ^{1}H NMR of (**4j**). Mg(OtBu)$_2$ (0.3 equiv), 60 °C, 1 h, Figure S13: ^{1}H NMR of (**4l**). Mg(OtBu)$_2$ (0.3 equiv), 60 °C, 1 h.

Author Contributions: M.P.: idea, structure, and design of the paper; planning for the work related to the publication; supervised analyzes of data; M.P. is the main supervisor. V.L. did the batch experiments and flow experiments, optimized most of the reaction conditions and analyzed the data. M.B. wrote the manuscript, is the supervisor of the bachelor student C.B., analyzed the data of C.B. C.B. performed the last optimization experiments and isolated products under the supervision of M.B. All authors have read, critically reviewed, and agreed to the final version of the manuscript.

Funding: This work was funded by the Université de Bordeaux and the CNRS. MB thanks the Ministère de l'Education nationale, de l'Enseignement supérieur et de la Recherche for a fellowship. V. Liautard thanks Aquitaine Sciences Transfer (AST) for funding.

Conflicts of Interest: The authors declare no conflict of interest.

References

1. Oppenauer, R.V. Eine Methode der Dehydrierung von Sekundären Alkoholen zu Ketonen. I. Zur Herstellung von Sterinketonen und Sexualhormonen. *Recl. Trav. Chim. Pays-Bas* **1937**, *56*, 137–144. [CrossRef]
2. Ponndorf, W.Z. Der reversible Austausch der Oxydationsstufen zwischen Aldehyden oder Ketonen einerseits und primären oder sekundären Alkoholen anderseits. *Angew. Chem.* **1926**, *39*, 138–143. [CrossRef]
3. Verley, M. A Simple and Efficient Catalytic Method for the Reduction of Ketones. *Bull. Soc. Chim. Fr.* **1925**, *37*, 871–874. [CrossRef]
4. Ramig, K.; Kuzemko, M.A.; Parrish, D.; Carpenter, B.K. A Novel Tandem Michael Addition/Meerwein −Ponndorf−Verley Reduction: Asymmetric Reduction of Acyclic α,β-Unsaturated Ketones Using A Chiral Mercapto Alcohol. *Tetrahedron Lett.* **1992**, *33*, 6279–6282. [CrossRef]
5. Wanner, M.J.; Koomen, G.J.; Pandit, U.K. A Facile Stereoselective Synthesis of (±)-Sesbanine. *Heterocycles* **1981**, *15*, 377–379. [CrossRef]
6. Bevinakatti, H.S.; Badiger, V.V.J. Synthesis and mass spectral studies of a coumarin analog of chloramphenicol. *Heterocycl. Chem.* **1980**, *17*, 1701–1703. [CrossRef]
7. Chinn, L.J.; Salamon, K.W. Synthesis of 9-deoxy-11-oxoprostaglandins. Selective reduction of an 11,15-dione. *J. Org. Chem.* **1979**, *44*, 168–172. [CrossRef]
8. Jones, R.A.; Webb, T.C. The chemistry of terpenes—V. *Tetrahedron* **1972**, *28*, 2877–2879. [CrossRef]
9. Rice, K.C.; Wilson, R.S. (-)-3-Isothujone, a small nonnitrogenous molecule with antinociceptive activity in mice. *J. Med. Chem.* **1976**, *19*, 1054–1057. [CrossRef] [PubMed]
10. de Graauw, C.F.; Peters, J.A.; van Bekkum, H.; Huskens, J. Meerwein-Ponndorf-Verley Reductions and Oppenauer Oxidations: An Integrated Approach. *J. Synthesis* **1994**, 1007–1017. [CrossRef]
11. Graves, C.R.; Joseph Campbell, E.; Nguyen, S.T. Aluminum-based catalysts for the asymmetric Meerwein-Schmidt-Ponndorf- Verley-Oppenauer (MSPVO) reaction manifold. *Tetrahedron Asymmetry* **2005**, *16*, 3460–3468. [CrossRef]
12. Graves, C.R.; Zeng, B.-S.; Nguyen, S.T. Efficient and Selective Al-Catalyzed Alcohol Oxidation via Oppenauer Chemistry. *J. Am. Chem. Soc.* **2006**, *128*, 12596–12597. [CrossRef] [PubMed]
13. Ooi, T.; Otsuka, H.; Miura, T.; Ichikawa, H.; Maruoka, K. Practical Oppenauer (OPP) Oxidation of Alcohols with a Modified Aluminum Catalyst. *Org. Lett.* **2002**, *4*, 2669–2672. [CrossRef] [PubMed]
14. Ooi, T.; Miura, T.; Itagaki, Y.; Ichikawa, H.; Maruoka, K. Catalytic Meerwein-Ponndorf-Verley (MPV) and Oppenauer (OPP) Reactions: Remarkable Acceleration of the Hydride Transfer by Powerful Bidentate Aluminum Alkoxides. *Synthesis* **2002**, 279–291. [CrossRef]

15. Gates, M.; Tschudi, G. The synthesis of Morphine. *JACS* **1956**, *78*, 1380. [CrossRef]
16. Woodward, R.B.; Wendler, N.E.; Brutschy, F.J. Quininone[1]. *JACS* **1945**, *67*, 1425. [CrossRef]
17. Rapoport, H.; Naumann, R.; Bissell, E.R.; Bonner, R.M. The preparation of some dihydro ketones in the morphine series by Oppenauer oxidation. *JOC* **1950**, *15*, 1103. [CrossRef]
18. Ballester, J.; Caminade, A.-M.; Majoral, J.-P.; Taillefer, M.; Ouali, A. Efficient and eco-compatible transition metal-free Oppenauer-type oxidation of alcohols. *Catal. Commun.* **2014**, *47*, 58–62. [CrossRef]
19. Kirihara, M.; Okada, T.; Sugiyama, Y.; Akiyoshi, M.; Matsunaga, T.; Kimura, Y. Sodium Hypochlorite Pentahydrate Crystals (NaOCl·5H$_2$O): A Convenient and Environmentally Benign Oxidant for Organic Synthesis. *Org. Process Res. Dev.* **2017**, *21*, 1925–1937. [CrossRef]
20. Almeida, M.L.S.; Beller, M.; Wang, G.Z.; Bäckvall, J.E. Chromium Complexes with Oxido and Corrolato Ligands: Metal-Based Redox Processes versus Ligand Non-Innocence. *Chem. A Eur. J.* **2018**, *2*, 1533–1536. [CrossRef]
21. Dani, P.; Karlen, T.; Gossage, R.A.; Gladiali, S.; van Koten, G. Hydrogen-Transfer Catalysis with Pincer-Aryl Ruthenium(II) Complexes. *Angew. Chem. Int. Ed.* **2000**, *39*, 743–745.
22. Hanasaka, F.; Fujita, K.; Yamaguchi, R. Cp*Ir Complexes Bearing N-Heterocyclic Carbene Ligands as Effective Catalysts for Oppenauer-Type Oxidation of Alcohols. *Organometallics* **2004**, *23*, 1490–1492. [CrossRef]
23. Ajjou, A.N.; Pinet, J.-L. Oppenauer-type oxidation of secondary alcohols catalyzed by homogeneous water-soluble complexes. *Can. J. Chem.* **2005**, *83*, 702–710. [CrossRef]
24. Coleman, M.G.; Brown, A.N.; Bolton, B.A.; Guan, H. Iron-Catalyzed Oppenauer-Type Oxidation of Alcohols. *Adv. Synth. Catal.* **2010**, *352*, 967–970. [CrossRef]
25. Battilocchio, C.; Hawkins, J.M.; Ley, S.V. A Mild and Efficient Flow Procedure for the Transfer Hydrogenation of Ketones and Aldehydes using Hydrous Zirconia. *Org. Lett.* **2013**, *15*, 2278–2281. [CrossRef] [PubMed]
26. Bruneau-Voisine, A.; Wang, D.; Dorcet, V.; Roisnel, T.; Darcel, C.; Sortais, J.-B. Transfer Hydrogenation of Carbonyl Derivatives Catalyzed by an Inexpensive Phosphine-Free Manganese Precatalyst. *Org. Lett.* **2017**, *19*, 3656–3659. [CrossRef] [PubMed]
27. Ogiwara, Y.; Ono, Y.; Sakai, N. Indium(III) Isopropoxide as a Hydrogen Transfer Catalyst for Conversion of Benzylic Alcohols into Aldehydes or Ketones via Oppenauer Oxidation. *Synthesis* **2016**, *48*, 4143–4148. [CrossRef]
28. Yohei, O.; Masahito, K.; Kotaro, K.; Takeo, K.; Norio, S. Oxidative Coupling of Terminal Alkynes with Aldehydes Leading to Alkynyl Ketones by Using Indium(III) Bromide. *Chem. A Eur. J.* **2015**, *21*, 18598–18600. [CrossRef]
29. Linghu, X.; Satterfield, A.D.; Johnson, J.S. Symbiotic Reagent Activation: Oppenauer Oxidation of Magnesium Alkoxides by Silylglyoxylates Triggers Second-Stage Aldolization. *J. Am. Chem. Soc.* **2006**, *128*, 9302–9303. [CrossRef] [PubMed]
30. Byrne, B.; Karras, M. Magnesium-oppenauer oxidation of alcohols to aldehydes and ketones. *Tetrahedron Lett.* **1987**, *28*, 769–772. [CrossRef]
31. Kloetzing, R.J.; Krasovskiy, A.; Knochel, P. The Mg-Oppenauer Oxidation as a Mild Method for the Synthesis of Aryl and Metallocenyl Ketones. *Chem. A Eur. J.* **2007**, *13*, 215–227. [CrossRef] [PubMed]
32. Knochel, P.; Millot, N.; Rodriguez, A.L.; Tucker, C.E. Preparation and Applications of Functionalized Organozinc Compounds. *Org. React.* **2004**. [CrossRef]
33. Klatt, T.; Markiewicz, J.T.; Sämann, C.; Knochel, P. Strategies To Prepare and Use Functionalized Organometallic Reagents. *J. Org. Chem.* **2014**, *79*, 4253–4269. [CrossRef] [PubMed]
34. Day, B.M.; Knowelden, W.; Coles, M.P. Synthetic and catalytic intermediates in a magnesium promoted Tishchenko reaction. *Dalt. Trans.* **2012**, *41*, 10930–10933. [CrossRef] [PubMed]
35. Fu, Y.; Zhao, X.L.; Hügel, H.; Huang, D.; Du, Z.; Wang, K.; Hu, Y. Magnesium salt promoted tandem nucleophilic addition–Oppenauer oxidation of aldehydes with organozinc reagent. *Org. Biomol. Chem.* **2016**, *14*, 9720–9724. [CrossRef] [PubMed]
36. Hallet, R.H.; Gries, G.; Gries, R.; Borden, J.H.; Czyzewska, E.; Oehlschlager, A.C.; Pierce, H.D.; Angerilli, N.P.D.; Rauf, A. Aggregation pheromones of two asian palm Weevils, Rhynchophorus ferrugineus and R. vulneratus. *Naturwissenschaften* **1993**, *80*, 328–331. [CrossRef]

37. Perez, A.L.; Hallett, R.H.; Gries, R.; Gries, G.; Oehlschlager, B. Pheromone chirality of asian palm weevils, Rhynchophorus ferrugineus (Oliv.) andR. vulneratus (Panz.) (Coleoptera: Curculionidae). *J. Chem. Ecology* **1996**, *22*, 357–368. [CrossRef] [PubMed]
38. Guerret, O.; Dufour, S. *Solid Composition for Controlled Delivery of Semiochemical Substances*; EP3187046A1; Melchior Material and Life Science: Lacq, France, 2017.

© 2018 by the authors. Licensee MDPI, Basel, Switzerland. This article is an open access article distributed under the terms and conditions of the Creative Commons Attribution (CC BY) license (http://creativecommons.org/licenses/by/4.0/).

Article

Prediction of In-Situ Gasification Chemical Looping Combustion Effects of Operating Conditions

Xiaojia Wang [1,*], Baosheng Jin [1], Hao Liu [2], Bo Zhang [1,*] and Yong Zhang [1]

1. Key Laboratory of Energy Thermal Conversion and Control of Ministry of Education, School of Energy and Environment, Southeast University, Nanjing 210096, China; bsjin@seu.edu.cn (B.J.); zyong@seu.edu.cn (Y.Z.)
2. Faculty of Engineering, University of Nottingham, Nottingham NG7 2RD, UK; liu.hao@nottingham.ac.uk
* Correspondence: xiaojiawang@seu.edu.cn (X.W.); bozhang@seu.edu.cn (B.Z.)

Received: 16 October 2018; Accepted: 5 November 2018; Published: 7 November 2018

Abstract: Chemical Looping Combustion (CLC) has been considered as one of the most promising technologies to implement CO_2 capture with low energy penalty. A comprehensive three-dimensional numerical model integrating gas–solid flow and reactions, based on the authors' previous work (Energy Fuels 2013, 27, 2173–2184), is applied to simulate the in-situ Gasification Chemical Looping Combustion (iG-CLC) process in a circulating fluidized bed (CFB) riser fuel reactor. Extending from the previous work, the present study further validates the model and investigates the effects of several important operating conditions, i.e., solids flux, steam flow and operating pressure, on the gas–solid flow behaviors, CO_2 concentration and fuel conversion, comprehensively. The simulated fuel reactor has a height of 5 m and an internal diameter of 60 mm. The simulated oxygen carrier is a Norwegian ilmenite and the simulated fuel is a Colombian bituminous coal. The results of this simulation work have shown that an increase in the solids flux can promote CO_2 concentration, but may also have a negative effect on carbon conversion. A decrease in the steam flow leads to positive effects on not only the CO_2 concentration but also the carbon conversion. However, the reduction of steam flow is limited by the CFB operation process. An increase in the operating pressure can improve both the CO_2 concentration and carbon conversion and therefore, the CFB riser fuel reactor of a practical iG-CLC system is recommended to be designed and operated under a certain pressurized conditions.

Keywords: chemical looping combustion; numerical prediction; CO_2 capture; fuel reactor; circulating fluidized bed

1. Introduction

Chemical Looping Combustion (CLC) is one of the most promising technologies to implement CO_2 capture with low energy penalty [1–3]. Since the introduction of CLC, numerous CLC studies on gaseous fuels have been conducted [4–11]. However, the more abundant reserves and higher carbon intensity of coal compared to gaseous fuels make the adoption of coal in the CLC system more attractive [12–17].

At present, the application of coal-fueled CLC mainly involves two possible options. The first approach is the so-called in-situ Gasification Chemical Looping Combustion (iG-CLC) which is usually based on the concept of Dual Interconnected Fluidized Bed (DIFB), as shown in Figure 1 [16,17]. Specifically, in the fuel reactor, the particles of the oxygen carrier are mixed with the coal particles and react with the gas products from coal pyrolysis and gasification. With a high conversion of combustion gases, the flue gas leaving the fuel reactor would mainly contain CO_2 and H_2O. For the particles (i.e., reduced oxygen carrier and unreacted char) entrained by the flue gas, they are directed into the separation system (including separators and a carbon stripper) to achieve the solid–solid and gas–solid separations. Through the separation process, the char particles are sent back to the fuel reactor while the particles of oxygen carrier are sent to the air reactor for re-oxidation. Thus, by virtue of the circulation of oxygen carrier particles to transport oxygen from the air reactor to the fuel reactor, the direct contact of coal and air can be avoided during the combustion process, and hence, highly purified CO_2 can be easily acquired at the outlet of the fuel reactor via the condensation of steam. The second possibility can be accomplished by a process integrating coal gasification and chemical looping combustion (IGCC-CLC) [18–21]. In this process, the syngas produced from coal gasification is used as the fuel of the downstream CLC system for power generation with CO_2 capture. Nevertheless, the defect of this approach is that the addition of a gasifier will increase system complexity and decrease efficiency. As mentioned by Spallina et al. [20] and Hamers et al. [21], a kind of packed-bed CLC reactor, which enables more operation flexibility and has no need for gas–solid separation, should be a potential candidate of this approach.

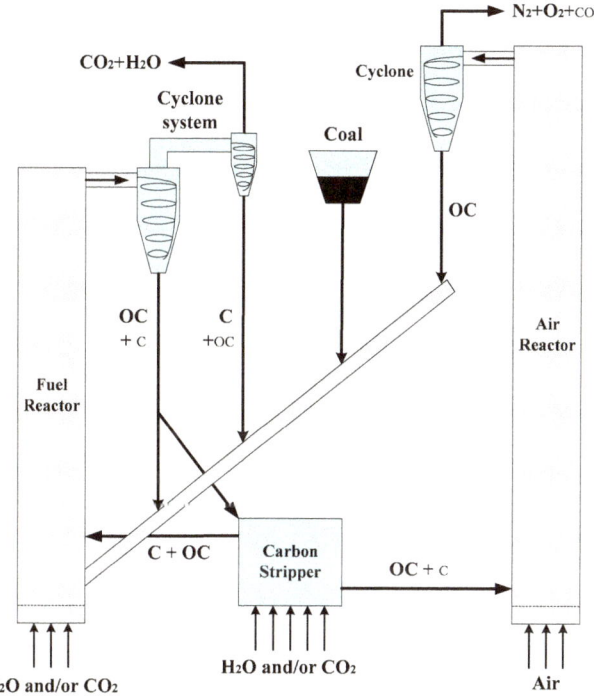

Figure 1. Schematic diagram of in-situ Gasification Chemical Looping Combustion system. Reprinted with permission from references [16,17]. Copyright 2013 Elsevier.

During the iG-CLC process, the oxygen carrier may suffer deactivation due to the presence of organic sulfur in the solid fuel and mass loss from elutriation, agglomeration and mixing with the coal ash [22,23]. Hence, it is very important to search for low-cost and long-lasting oxygen carriers. In this respect, natural iron ores, owing to their low-cost, adequate oxygen transport capacity, favorable reactivity and environmental friendliness, together with good mechanical stability, have been regarded as one of the most feasible oxygen carrier candidates for the iG-CLC [24–28].

Currently, the iG-CLC approach has been investigated in some pilot-scale units with different reactor designs, preliminarily demonstrating the feasibility and potential of iG-CLC [16,18,29–36]. However, in view of the gathered results, there are still several challenges affecting the development of this technology: (1) The existence of unburned gaseous compounds at the outlet of fuel reactor results in unsatisfactory combustion efficiency, and CO_2 capture concentration [37–39]. (2) The conversion of char is unfavorable due to the slow gasification rate under the operating temperatures of the fuel reactor (usually 800–950 °C). Thus, a part of unreacted char particles off the fuel reactor may enter and burn in the air reactor, which will lead to the reduction of carbon capture efficiency [39–41]. It can be found that both of the two issues are directly related with the fuel reactor. Hence, a feasible iG-CLC system should have a good selection of its fuel reactor type/structure which can promise smooth operations and efficient fuel conversions. In this respect, circulating fluidized bed (CFB) reactors, owing to their inherent advantages in terms of favorable gas–solid contacts over the whole reactor height with a relatively lower solids inventory, are becoming a competitive candidate for the fuel reactor of iG-CLC [16,17,31–33,35,36]. A deep understanding of the effects of operating conditions on the flow and reaction performance of the CFB fuel reactor allows for the optimization of the system operations to achieve the best whole-system fuel conversion, and the highest CO_2 concentration and carbon capture efficiency at the outlet of the fuel reactor. However, some important operational information, such as the effects of operating conditions on instantaneous flow behaviors and heterogeneous reaction characteristics, is still largely missing and rarely available for the iG-CLC process with a CFB fuel reactor.

With the advancement of computer hardware and numerical methods, Computational Fluid Dynamics (CFD) modeling is now able to predict fairly accurate gas–solid flow behaviors and chemical reactions. Therefore, it has been accepted as a reliable technique to make up for the disadvantages of experimental methods (e.g., instantaneous gas–solid hydrodynamics and reaction dynamics), and further give some forward guidance for deeper experimental studies [42–47]. Previous publications on the simulation studies of CLC mainly focused on CLC of gaseous fuels [48–53] and few were on the CLC performance of solid fuels. Mahalatkar et al. [54] spearheaded a 2D model to simulate the flow and reaction process in a bubbling fluidized bed (BFB) fuel reactor. Su et al. [55] simulated the iG-CLC process in a dual circulation fluidized bed system, where the gas leakage, flow pattern and combustion efficiency were obtained and analyzed. Shao et al. [56] developed a 3D full-loop iG-CLC model which successfully predicted some important flow behaviors of the whole system. However, the coupling of the reaction model into the hydrodynamics model had not been achieved in their work, and hence, only cold-state flow characteristics were presented. Alobaid et al. [57] and May et al. [58] developed 3D models for fuel and/or air reactor of the world's second largest CLC pilot plant at Technische Universität Darmstadt, enabling scale-up investigations for the optimization of process and solid–fluid interactions. Besides, an in-house thermochemical reaction model was successfully combined with the Euler–Euler model in their work. Generally speaking, by far, the simulation studies on iG-CLC technology mainly concentrated on the model feasibility tests, but the in-depth investigations of flow and reaction mechanisms in the fuel reactor are still lacking.

In our previous work, we successfully developed a comprehensive 3D numerical model of a CFB fuel reactor based upon the experimental system of others [59] to study the fundamental performance of iG-CLC, including fuel conversions and reaction rates [22]. With the present work, aiming to improving the existing problems, in terms of insufficient combustion efficiency, CO_2 concentration, and carbon capture efficiency, in the iG-CLC process, we have conducted in-depth simulations of the effects of the important operating conditions, i.e., solids flux, steam flow and operating pressure, on the iG-CLC performance. The main contributions of this simulation work are listed as follows: (1) further validation of a three-dimensional numerical model, including the kinetic theory of granular flow and complicated heterogeneous reactions, for the simulation of the iG-CLC system with a CFB as the fuel reactor; (2) predictions of detailed flow dynamic characteristics in the fuel reactor under different hot-state conditions; (3) in-depth study of the effects of important operating parameters on the reaction performance by integrating the gas–solid flow mechanisms; (4) effective complement of the limitations of current experimental conditions on flow and reaction mechanisms, and further offering valuable guidance for the design and operation of future iG-CLC plants.

2. Results and Discussion

Table 1 lists the main operating parameters in this simulation work, except for the three variables to be mentioned below.

Table 1. Main operating parameters used in simulations.

Parameters	Value
Reactor diameter (mm)	60
Reactor height (m)	5
Coal feeding rate (kg/h)	10
Thermal Power (kW$_{th}$)	~60
Reactor temperature (K)	1273

Nine different cases were simulated to study the effects of operating conditions including solids flux, steam flow and operating pressure on the flow behavior and reaction performance in the fuel reactor. Here, CO_2 concentration and single-loop conversion of carbon are the two main research objects to reflect the reaction performance. A high CO_2 concentration at the exit of the fuel reactor means high combustion efficiency with efficient CO_2 capture. The higher the single-loop conversion of carbon in the fuel reactor, the higher the carbon capture efficiency and the thermal power that could be achieved. In order to investigate the effect of each parameter on the reaction performance more rationally, all other parameters of the cases to be compared had been kept constant when the parameter to be investigated was changed. Table 2 lists the nine simulation cases with the specified operating conditions, where case 1 is selected as the reference condition for the analysis of basic flow and reaction behaviors.

Table 2. Simulation cases and operating conditions.

Case	Solids Flux, G_p (kg/m^2·s)	Steam Flow, Q_h (m^3/h)	Operating Pressure, P_0 (MPa)
1–3	100–200–300	7.5	0.55
4–6	220	6 8 10	0.6
7–9	200	7	0.4–0.6–0.8

2.1. Further Validation of the Hydrodynamics and Reaction Models

Before carrying out the comprehensive modeling of iG-CLC, a further validation of the hydrodynamics model was carried out by comparing the model predictions with the experimental data of a different CFB [60] from that used for validation of hydrodynamics model with our previous study [22,59]. The dimensions of the modeled CFB were the same as those of the experimental facility used by Jin et al. [60], except for a small difference in the exit geometry. The simulation parameters were also set accordingly to their experimental conditions. Figure 2 shows the comparison between the computed and experimental axial pressure gradient profiles. The modeling results were very close to the experimental data, and the maximum relative error between the simulation and the experiment was less than 12%. Additionally, our calculations were also close to the previous simulation predictions by Jin et al. [60]. Hence, the hydrodynamics model adopted in this study can be considered suitable for the simulation of gas–solid flow in the CFB fuel reactor.

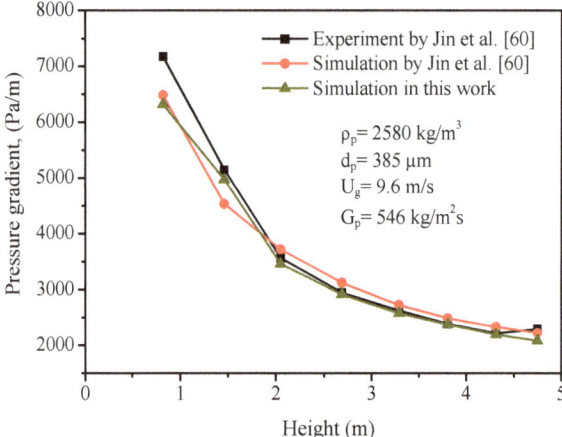

Figure 2. Comparisons of axial pressure gradient profiles between the experiment and simulation.

In order to verify the reaction model adopted in this study, we carried out a comparison of reaction performance in a BFB fuel reactor between the simulation using our reaction model and the previous experimental data [28]. The simulation conditions were set to be consistent with the experimental operation conditions (temperature-1163 K, the steam flow-190 L_N/h, and the coal feeding flow-62 g/h). Figure 3 shows the comparison of dry basis gas concentrations at the fuel reactor outlet between the experimental data and our simulations. It shows that the simulation results were generally in agreement with the experimental data, indicating that the reaction model is valid. A similar validation of the reaction model using the experimental data obtained under different conditions (temperature-1163 K, the steam flow-190 L_N/h, and the coal feeding flow-83 g/h) from the same literature [28] had also been performed in our previous work [22].

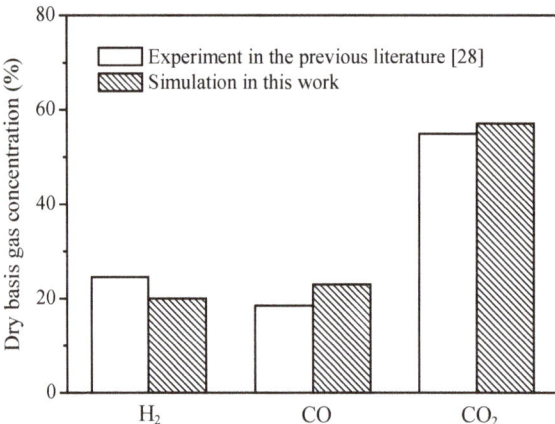

Figure 3. Comparison of dry basis gas concentrations at the fuel reactor outlet between the experiment and simulation.

From the above, we can conclude that both the hydrodynamics model and the reaction model which make up the comprehensive model of this study have been further validated in advance of the present simulation studies. Therefore, the simulated results on the gas–solid flow patterns and reaction performance in the coal-fired CFB fuel reactor using the comprehensive model can be considered to be valid.

2.2. Distributions of Solids Holdup and Gas–solid Components

Figure 4a displays the distribution of solids holdup with the reactor height at the quasi-equilibrium state (here, the time of 70 s was selected as a representative) under the reference condition. To enhance the clarity of presentation, the colored contours of the solids holdup in three 0.1 m-long sections of the reactor at different height levels (0.5–0.6 m, 2.5–2.6 m, 4.5–4.6 m) are shown in Figure 4a. Similar approaches have been adopted for other colored contours in Figure 4b–f. It can be seen that the solids holdup keeps decreasing along with the bed height in general, except a slight increase near the top of riser due to the effect of the L-shaped exit configuration [61]. The volume averages of solids holdup and axial solid velocity in the fuel reactor are about 0.07 and 0.70 m/s, respectively. Accordingly, the solids inventory and residence time in the fuel reactor can be calculated as 60.3 kg/MW$_{th}$ and 7.1 s, respectively.

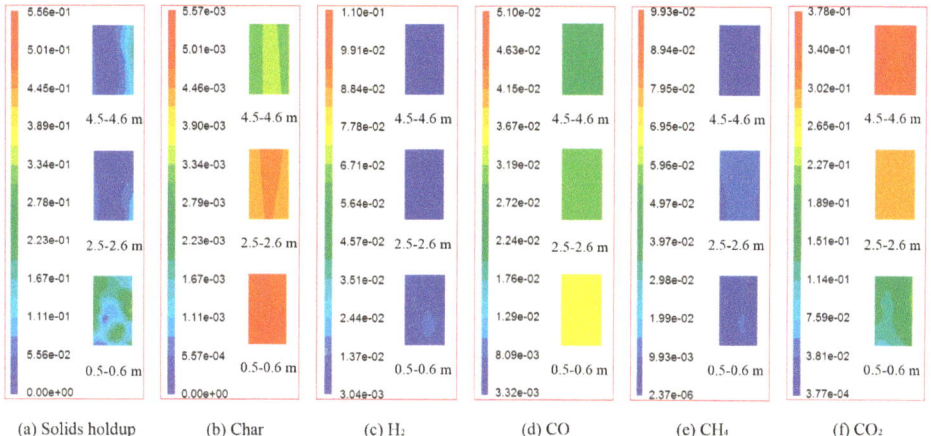

Figure 4. Axial (x = 0) distributions of solids holdup and gas–solid components along the bed height at 70 s with the reference condition: G_p = 100 kg/m^2·s, Q_h = 7.5 m^3/h, P_0 = 0.55 MPa. (**a**) solids holdup; (**b**) mass fraction of char in solid phase; (**c**–**f**) molar fractions of gas species in gas phase.

The patterns of the char mass fraction in the solid phase and the gas concentrations in the gas phase at 70 s under the reference condition are presented in Figure 4b–f. At the bottom of the reactor, the rapid reaction of coal pyrolysis leads to the generation of char in abundance so that the mass fraction of char in the solid phase reaches the peak in this section. Then, the mass fraction of char begins to decrease along with the reactor height due to the continuing consumption of char from the gasification reactions. Similarly, H_2 is generated from coal pyrolysis and gasification, but is quickly consumed through the combustion reactions of intermediate gasification products and oxygen carrier. Thus, we could observe a decreasing concentration of H_2 along with the reactor height. The trends of CO and CH_4 are consistent with that of H_2. Figure 4f presents the distribution of CO_2 concentration in the fuel reactor. As the product gas, the distribution of CO_2 is almost opposite to that of reactant gases (i.e., H_2, CO and CH_4). The more detailed discussions about the flow patterns and component distributions in the bed could be found in our previous preliminary study [22].

It should be noted that the single-loop conversion of carbon and CO_2 dry-basis concentration at the outlet under the reference condition reach 65.6% and 90.2%, respectively, which are comparable to the previous experimental results by Berguerand and Lyngfelt [12]. In their 10 kW$_{th}$ system, the single-loop conversion of carbon was in the range of 50–80% while the CO_2 dry-basis concentration ranged between 78% and 81%. This indicates the CFB riser with high solids flux is a potential candidate for the fuel reactor of iG-CLC [23,36,62], and the operating conditions adopted in this study are appropriate and feasible.

2.3. Effect of Solids Flux

Because a higher solids flux can enhance the solids holdup and heat carrying capacity, CFBs with high solids mass fluxes ($G_p \geq 200$ kg/m^2·s) are considered as promising reactors for some special processes such as the production of maleic anhydride, combustion and gasification [63–65]. In the future commercial operation of iG-CLC, it is more important to achieve high CO$_2$ concentration than to acquire low solids inventories when the cheap oxygen carriers with a relatively low reactivity are used. Hence, it is very interesting to study the iG-CLC performance under the condition of a high solids flux. Simulations with Cases 1–3 have been performed with different solids fluxes ranging from 100 kg/m^2·s to 300 kg/m^2·s to investigate the effect of solids flux on the performance of iG-CLC when the steam flow and operating pressure are kept at constant values of 7.5 m^3/h (under the operating temperature and pressure) and 0.55 MPa, respectively. In these cases, the steam to the fixed carbon ratio (H$_2$O/C) is kept at the value of 0.85. With the coal feeding rate being kept at a constant value of 10 kg/h, changing the solids flux from 100 kg/m^2·s to 300 kg/m^2·s results in a change of the mass flow ratio of oxygen carrier to coal from about 101.7–305.2.

Figure 5 shows the apparent solids holdups (i.e., volumetric solids fraction) along the fuel reactor height as a function of solids flux. It can be seen that the range of solids holdup is much wider in the lower region than that in the upper region at a given solids flux, implying that the particle concentration distribution is more uniform in the upper dilute region. Meanwhile, we can observe a significant increase of solids holdup with the increasing solids flux. Moreover, the increase of solids flux promotes the realization of a fully/largely high-density CFB ($\alpha_p \geq 0.1$). When the solids flux increases from 100–300 kg/m^2·s, the height of the high-density solids flow region increases from 0.9 m to about 1.8 m. These computational results are consistent with the experimental results of several groups of researchers [22,23,61,66,67].

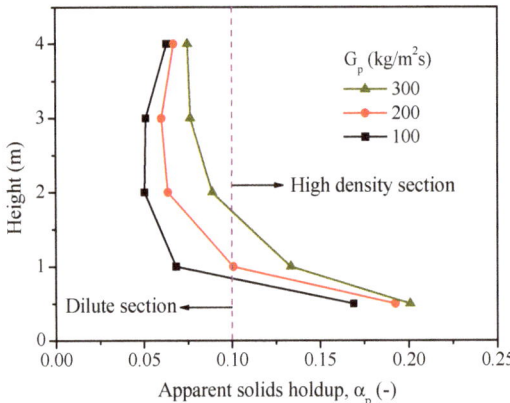

Figure 5. Mean solids holdups in different heights as a function of solids flux.

Figure 6 presents the CO$_2$ concentration (dry basis) at the outlet vs. time curves with different solids fluxes. The concentration of CO$_2$ at a given solids flux has a slight oscillation around a constant value after reaching the quasi-equilibrium state ($t \geq 30$ s). We can observe that the increase of solids flux results in an increase in the CO$_2$ concentration at the fuel reactor outlet. The average dry basis concentration of CO$_2$ is about 90.2% for the solids flux of 100 kg/m^2·s and increases to 96.3% for the solids flux of 300 kg/m^2·s. The increase in the CO$_2$ concentration at the outlet of the reactor is believed to be the results of higher solids holdups (Figure 5) and better gas–solid contacts which promote the gas–solid reactions, according to reactions (8)–(10) (see Section 3.2).

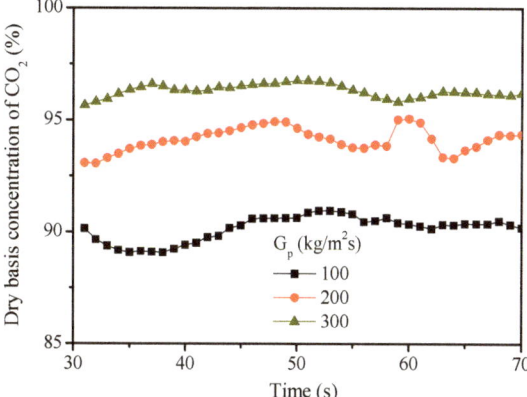

Figure 6. Effect of solids flux on the CO$_2$ concentration (dry basis) at the outlet of the fuel reactor.

Figure 7 shows the profiles of the mean axial velocities of the solids along the reactor height with different solids fluxes. It can be observed that the particles are generally accelerating along with the riser, which is largely due to the drive of gases. At the same time, the axial solids velocities increase with the solids flux. This can be explained by the fact that the gas phase is accelerated with higher solids fluxes and holdups, which further leads to higher particle velocities. These modeling results are consistent with the experimental observations of Pärssinen and Zhu [59].

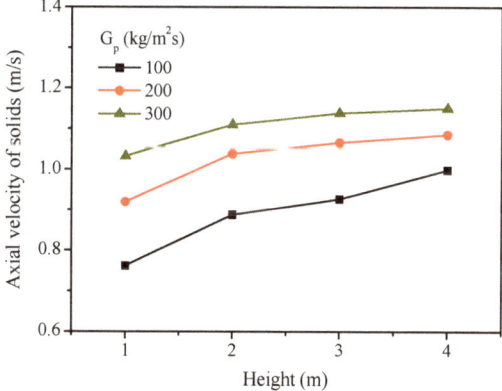

Figure 7. Distribution of mean axial solids velocities as a function of height with different solids fluxes.

Figure 8 shows the effect of the solids flux on the single-loop conversion of carbon in the fuel reactor. The conversion of carbon decreases from 65.6% to 50.1% with the increase of solids flux from 100 kg/m^2·s to 300 kg/m^2·s. As mentioned above, the increase in the solids velocity with a higher solids flux would result in a decrease in the solids residence time, which leads to a reduction of char conversion. On the other hand, according to reactions (6)–(7) (see Section 3.2), the increases in the fuel gas conversion and CO$_2$ concentration with a higher solids flux will promote the gasification of char, i.e., the conversion of char. Hence, the net effect of the solids flux on the single-loop conversion of carbon is a combination of the positive (due to the increased gasification reactions (6)–(7) and the negative (due to the reduced solids residence time) effects. In the end, under the modeling conditions of this study, higher solids fluxes have been found to have some negative effects on the carbon conversion, indicating that the change of solids velocity with different solids fluxes may play a leading role in the carbon conversion.

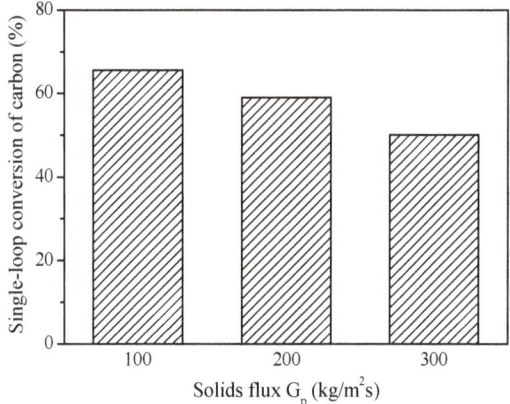

Figure 8. Effect of solids flux on the single-loop conversion of carbon in the fuel reactor.

2.4. Effect of Steam Flow

Steam (H_2O) plays a very important role in the iG-CLC process. It is not only the fluidizing gas but also the main gasification agent. Hence, it is necessary to understand the effect of steam flow on the iG-CLC performance. Cases 4–6 are designed to study the influence of steam flow on the flow patterns and reaction characteristics. The steam flow selected is in the range of 6–10 m^3/h (under the operating temperature and pressure) with the solids flux, coal feeding rate and operating pressure being fixed at values of 220 kg/m^2·s, 10 kg/h and 0.6 MPa, respectively. Thus, in these cases, the steam to the fixed carbon ratio (H_2O/C) ranges from 0.74 to 1.23, whereas the mass flow ratio of the oxygen carrier to the coal feeding rate is kept at a constant value of 223.8.

Figure 9 presents the profiles of the apparent solids holdups along the fuel reactor height as a function of steam flow. The solids holdup generally increases with the decreasing steam flow at a given height [23], and the variation of solids holdup is smaller in the higher riser region. These results illustrate that a higher steam flow has a negative effect on the formation of high-density flow structure, which agrees with the experimental observations of Wang et al. [23], Wang et al. [61] and Li et al. [66].

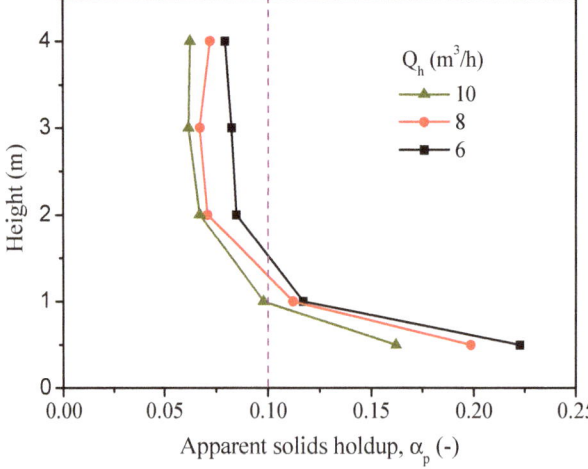

Figure 9. Mean solids holdups as a function of steam flow in different heights.

Figure 10 shows the mean axial velocities of solids with different steam flows as a function of the fuel reactor height. As expected, the axial velocity of the solids decreases with the decreasing steam flow at a specified height. The axial velocity profiles of Wang et al. [61] and Pärssinen and Zhu [59] agree well with the results of our simulation but with higher particle velocities due to the higher gas velocities used in their studies.

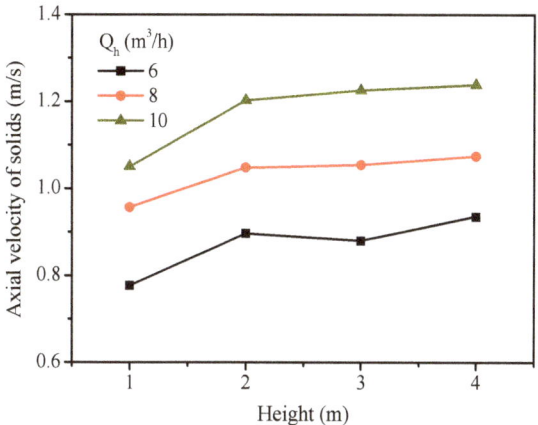

Figure 10. Solids axial velocities in different steam flow as a function of height.

Figure 11 shows the effect of the steam flow on the CO_2 concentration (dry basis) at the outlet of the fuel reactor. Decreasing the steam flow gives rise to an increase in CO_2 concentration at the outlet. The average concentration of CO_2 (dry basis) is 93.7% for the steam flow of 10 m^3/h and increases to 96.1% when the steam flow is decreased to 6 m^3/h. The increase of CO_2 concentration with a reduced steam flow is believed to be related to the higher solids holdup and the lower gas velocity, which further enhance the gas–solid contacts and the residence time, promoting gas–solid reactions (8)–(10).

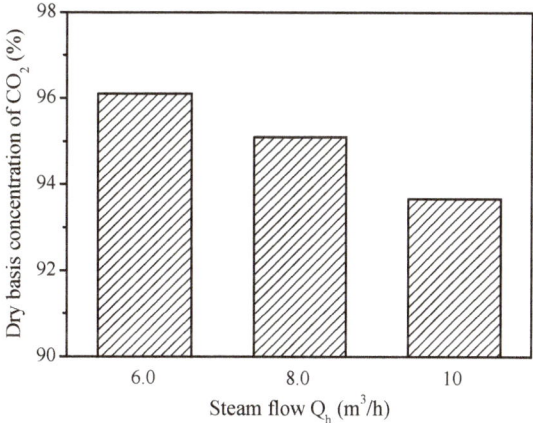

Figure 11. Effect of steam flow on the CO_2 concentration (dry basis) at the outlet of the fuel reactor.

Figure 12 presents the influence of steam flow on the single-loop carbon conversion in the fuel reactor. The carbon conversion increases from 56.0% to 65.3% when the steam flow decreases from 10 m^3/h to 6 m^3/h. When the steam flow is reduced, the solid residence time is increased, and the char gasification can be enhanced due to the increases in the fuel gas conversion and CO_2 concentration (according to reactions (6)–(7)), leading to an increase in carbon conversion.

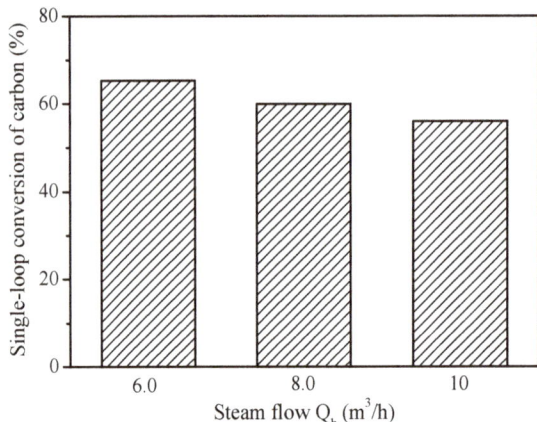

Figure 12. Effect of steam flow on the single-loop conversion of carbon in the fuel reactor.

However, although a decrease in the steam flow can be beneficial to the CO_2 concentration at the exit and the carbon conversion, lower steam flow would go against the rapid circulation of the particles, which would lead to an increase in the solid inventory and a reduction in the heat transfer efficiency of the oxygen carrier. Moreover, too low steam flow cannot even drive the particle flow for circulating. Hence, in order to ensure the proper circulation and fluidization of the particles, the reduction in the steam flow needs to be limited by the operation process.

2.5. Effect of Operating Pressure

Because coal gasification is the rate-controlling step in the process of iG-CLC, an increase in the partial pressure of the gasification agent can have a significant influence on the performance of iG-CLC [13,68]. Meanwhile, in a pressurized CLC system, electric power can be saved to compress the CO_2 flow to the high pressure needed for transportation and storage [7,51]. Hence, a study on the pressurized iG-CLC performance is necessary. Simulations with Cases 7–9 have been performed with different operating pressures ranging from 0.4 MPa to 0.8 MPa to investigate the effect of the operating pressure on the gas–solid flow and reaction behaviors with the solids flux, coal feeding rate and steam flow being kept at constant values of 200 kg/m²·s, 10 kg/h and 7 m³/h (under the operating temperature and pressure), respectively. In these cases, the steam to the fixed carbon ratio (H_2O/C) is in the range of 0.14–1.15, whereas the mass flow ratio of the oxygen carrier to the coal feeding rate is kept at a constant value of 203.5.

Figure 13 shows the effect of the operating pressure on the distribution profiles of the apparent solids holdups along the fuel reactor height. Generally, the solids holdup goes up with the increasing operating pressure at a given height. An increase in the operating pressure also leads to an increase in the height of the high-density flow region, indicating the active effect of pressure on the forming of a fully high-density CFB [69]. Figure 14 displays the mean axial velocities of solids with different operating pressures as a function of height. It can be observed that the axial solid velocity decreases with the increasing pressure at a specified height.

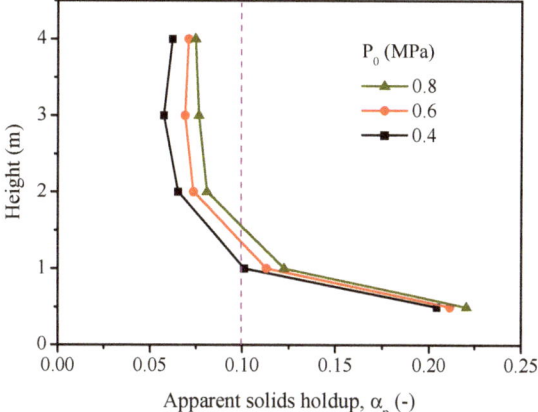

Figure 13. Effect of operating pressure on the profile of the solids holdups along the reactor height.

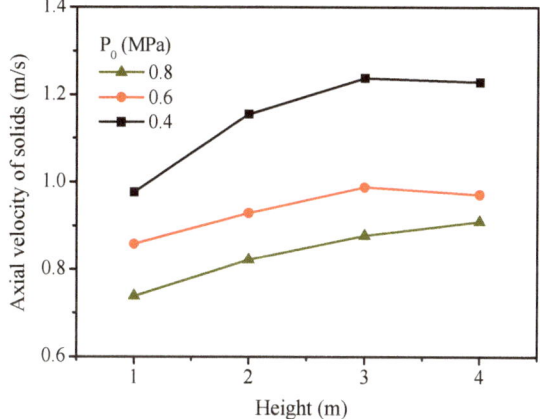

Figure 14. Distribution of axial velocities of solids with different pressures as a function of height.

Figure 15 presents the influence of pressure on the CO_2 concentration (dry basis) at the outlet of the fuel reactor. An increase in operating pressure leads to an increase in CO_2 concentration. The average concentration of CO_2 (dry basis) increases from 93.9% to 95.4% when the pressure rises from 0.4 MPa to 0.8 MPa. According to the chemical reactions of iG-CLC, particularly, reactions (8)–(10), the increase of CO_2 concentration under a higher pressure condition may be attributed to the following three factors. Firstly, the concentrations of gasification intermediates are increased at elevated pressures, and this promotes the reduction reactions (8)–(10) with the oxygen carrier. Secondly, as shown in Figure 13, the increased operating pressure augments the solids holdup, which further enhances the gas–solid contact and promotes gas–solid reactions. Finally, the gas residence time is longer in the pressurized condition [13,68].

Figure 16 shows the influence of the operating pressure on the single-loop conversion of carbon in the fuel reactor. The carbon conversion goes up substantially from 48.4% to 70.8% with the operating pressure increasing from 0.4 MPa to 0.8 MPa. The elevated steam partial pressure and longer gas residence time facilitate char gasification according to reactions (6)–(7) [68]. Moreover, the elevated pressure leads to a decrease in solids velocity, thus, longer solid residence time and higher char conversion. The predicted results shown in Figure 16 are consistent with the experimental observations of Xiao et al. [26] who conducted their experiments with a fixed bed reactor.

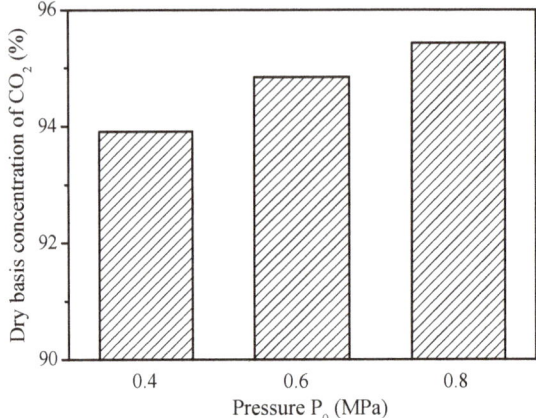

Figure 15. Effect of pressure on the CO₂ concentration (dry basis) at the outlet of the fuel reactor.

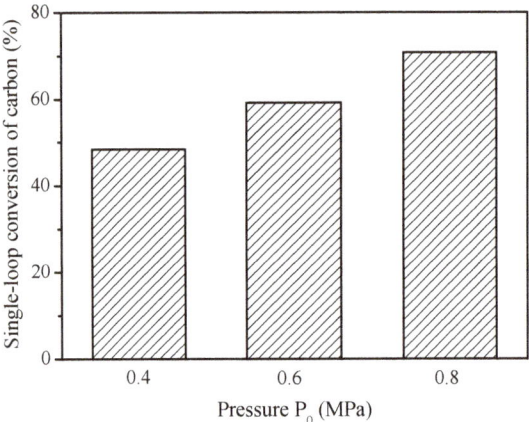

Figure 16. Influence of operating pressure on the single-loop carbon conversion in the fuel reactor.

Several researchers (e.g., García-Labiano et al. [7]) found the negative effect of the increasing total pressure on the reduction reaction performance of oxygen carrier. However, for iG-CLC, the rate-controlling step of the whole reaction process is the gasification reactions (6)–(7) of char instead of the subsequent reduction reactions of oxygen carrier [22,39–41]. In addition, the possible negative effect for the ilmenite adopted in this study has not been mentioned in the experimental research by Abad et al. [24] and Cuadrat et al. [70] Therefore, the possible negative effect of high operating pressures on the reduction reactions of the oxygen carrier was neglected in this study.

3. Materials and Methods

The comprehensive hydrodynamics and reaction models adopted in this study are based on our previous study [22]. But in order to carry out further and independent validations, we selected a new CFB fuel reactor with a different bed height and diameter from that simulated previously [22,59]. As shown in Figure 17, the fuel reactor simulated in this study is a circulating fluidized bed (CFB) riser with a height of 5 m and an internal diameter of 60 mm by reference to the previous experiments [60,61].

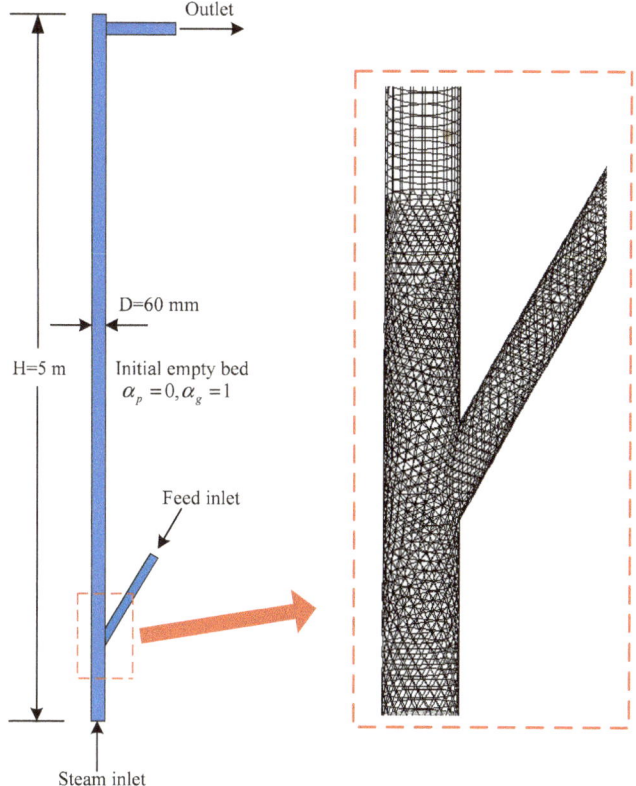

Figure 17. Sketch and grids of the CFB fuel reactor.

3.1. Governing Equations

In this study, a Euler–Euler model is used to simulate the hydrodynamics of gas–solid flow in the fuel reactor, coupled with chemical reaction models. The standard k–ε model is applied to simulate the gas phase turbulence and the kinetic theory of granular flow (KTGF) is adopted to simulate the solid phase. The mass, momentum and energy conservation equations are applied to both gas and solid phases. The species conservation equations are solved for individual species in each phase. Here, the conservation equations for gas phase (i.e., Equations (1)–(3)) and the species transport equation for gas species (i.e., Equation (4)) are given as the representatives [22,47]. Meanings of the symbols can be found in the nomenclature.

$$\frac{\partial}{\partial t}\left(\alpha_g \rho_g\right) + \nabla \cdot \left(\alpha_g \rho_g \vec{v}_g\right) = \dot{m}_{gp} \tag{1}$$

$$\frac{\partial}{\partial t}\left(\alpha_g \rho_g \vec{v}_g\right) + \nabla \cdot \left(\alpha_g \rho_g \vec{v}_g \vec{v}_g\right) = -\alpha_g \nabla p_g + \nabla \cdot \bar{\bar{\tau}}_g + \alpha_g \rho_g \vec{g} + \dot{m}_{gp} \vec{v}_p + \beta(\vec{v}_p - \vec{v}_g) \tag{2}$$

$$\frac{\partial}{\partial t}\left(\alpha_g \rho_g H_g\right) + \nabla \cdot \left(\alpha_g \rho_g \vec{v}_g H_g\right) = \nabla \cdot \left(\lambda_g \nabla T_g\right) + h_{gp}(T_g - T_p) + \dot{m}_{gp} H_p \tag{3}$$

$$\frac{\partial}{\partial t}\left(\alpha_g \rho_g Y_{g,i}\right) + \nabla \cdot \left(\alpha_g \rho_g Y_{g,i} \vec{v}_g\right) = -\nabla \cdot \alpha_g J_{g,i} + S_i \tag{4}$$

The closure models applied in this simulation to describe constitutive relations are given in Table 3.

Table 3. Closure models used in simulations.

Description	Model	References
Turbulence model	standard k–ε model	[60,71]
Drag model	Gidaspow	[72]
Granular viscosity	Gidaspow	[72]
Granular bulk viscosity	Lun et al.	[73]
Granular temperature	Algebraic	[51]
Solids pressure	Lun et al.	[73]
Radial distribution	Lun et al.	[73]

In this study, considering the small bed temperature variation along the whole reactor height and the excellent performance of interphase heat transfer in the circulating fluidized bed, the reactor is assumed to be isothermal and the differences of temperature between gas–solid phases are ignored for simplification. The detailed descriptions of governing equations can be found in our previous publication [22].

3.2. Chemical Reactions

As part of the comprehensive numerical model, the complex homogeneous and heterogeneous chemical reactions are considered and coupled into the solver by setting the source terms of continuity, momentum and species transport equations. On this basis, the consumption of the reactants and the harvest of the products during the reaction processes can be calculated in real time [49].

The reactions occurring in the fuel reactor involve the coal pyrolysis (5), the gasification of char (6) and (7), and the oxidation reactions of the intermediate gasification products with the oxygen carrier (8)–(10). All these reactions are included by user defined functions (UDF). The detailed reaction steps are summarized below.

First, coal pyrolysis occurs through reaction (5):

$$\text{Coal(s)} \rightarrow a\text{CO(g)} + b\text{H}_2\text{(g)} + c\text{CH}_4\text{(g)} + d\text{CO}_2\text{(g)} + e\text{H}_2\text{O(g)} + f\text{Char(s)} \tag{5}$$

The char is gasified with H_2O and CO_2 by reactions (6)–(7):

$$\text{C(s)} + \text{H}_2\text{O(g)} \rightarrow \text{CO(g)} + \text{H}_2\text{(g)} \quad \Delta H^\theta_{1273\ K} = 135.6\ \text{kJ/mol} \tag{6}$$

$$\text{C(s)} + \text{CO}_2\text{(g)} \rightarrow 2\text{CO(g)} \quad \Delta H^\theta_{1273\ K} = 167.8\ \text{kJ/mol} \tag{7}$$

As shown in Reaction (5), the volatile matter in coal was assumed to be released in the form of H_2, CO, CH_4 and CO_2. Other small amounts of light hydrocarbons and tars in the volatiles were not considered in this study [22,70].

The main active components of the Norwegian ilmenite, adopted as the oxygen carrier in this study, include Fe_2TiO_5 and Fe_2O_3. Here, the Integrated Rate of Reduction (IRoR) model was adopted, with which the Fe_2TiO_5 was considered as a mixture of Fe_2O_3 and TiO_2; $FeTiO_3$, a mixture of FeO and TiO_2; Fe_3O_4, a mixture of Fe_2O_3 and FeO; and TiO_2, as an inert material [22,24,74]. Thus, the reduction reactions of ilmenite can be simplified as the following reactions (8)–(10):

$$4\text{Fe}_2\text{O}_3\text{(s)} + \text{CH}_4\text{(g)} \rightarrow 8\text{FeO(s)} + \text{CO}_2\text{(g)} + 2\text{H}_2\text{O(g)} \quad \Delta H^\theta_{1273\ K} = 268.9\ \text{kJ/mol} \tag{8}$$

$$\text{Fe}_2\text{O}_3\text{(s)} + \text{H}_2\text{(g)} \rightarrow 2\text{FeO(s)} + \text{H}_2\text{O(g)} \quad \Delta H^\theta_{1273\ K} = 18.4\ \text{kJ/mol} \tag{9}$$

$$\text{Fe}_2\text{O}_3\text{(s)} + \text{CO(g)} \rightarrow 2\text{FeO(s)} + \text{CO}_2\text{(g)} \quad \Delta H^\theta_{1273\ K} = -13.7\ \text{kJ/mol} \tag{10}$$

The reaction rates are represented by the Arrhenius law with the pre-exponential factors and activation energies obtained from the previous literatures [25,70,75]. Further details on the reaction kinetics and the model can be found from our previous publication [22].

Figure 18 shows the standard Gibbs free energy changes for the main reactions (6)–(10) in the fuel reactor calculated over a wide range of operating temperatures. It can be seen that the Gibbs free energy decreases with an increase in the reaction temperature for all of these reactions. As the reaction affinity is enhanced with a decrease in the Gibbs free energy [76], high temperatures favor the reactions in the fuel reactor. Therefore, we selected a relatively high temperature of 1273 K as the temperature of the fuel reactor to be simulated.

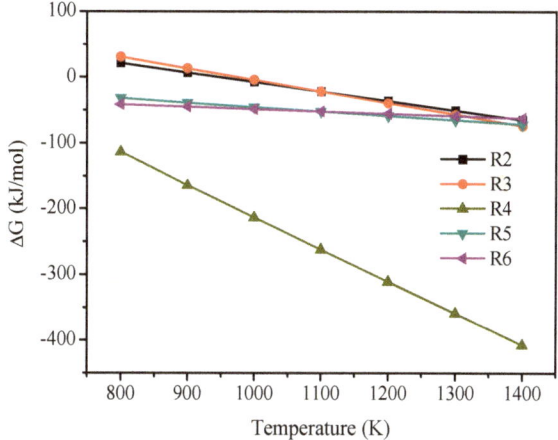

Figure 18. Standard Gibbs free energy changes for reactions in the fuel reactor.

3.3. Data Evaluation

The dry basis component concentration (f_i) in the exhaust gas is calculated as follows [22]:

$$f_i = \frac{x_i}{x_{CO} + x_{CO_2} + x_{CH_4} + x_{H_2}} \times 100\% \tag{5}$$

where x_i is the molar fraction of species i (e.g., CO, CO_2, CH_4 or H_2) in the gas phase.

The single-loop conversion of carbon (X_C) is calculated as [22]:

$$X_C = (1 - \frac{Q_{C,out}}{Q_{Coal,in} Y_C}) \times 100\% \tag{6}$$

where Y_C is the mass fraction of carbon in the original coal, $Q_{Coal,in}$ is the mass flux of coal at the solids inlet and $Q_{C,out}$ is the mass flux of unreacted char at the outlet.

The solids flux (i.e., solid circulation rate) G_p is calculated as:

$$G_p = U_{p,f} \alpha_{p,f} \rho_p \frac{A_f}{A_s} \tag{7}$$

where $U_{p,f}$ is the particle velocity at the feed inlet, $\alpha_{p,f}$ is the particle volume fraction at the feed inlet, A_f is the sectional area of the feed inlet, and A_s is the sectional area of the steam inlet (i.e., the sectional area of the fuel reactor).

3.4. Numerical Considerations

In this study, simulations were carried out in three-dimensional (3D) domains. The pressure-based solver was employed for solving the governing equations. The phase-coupled semi-implicit method for pressure-linked equations (PC-SIMPLEs) algorithm was applied to deal with the pressure–velocity coupling. After mesh independence analyses, the hexahedral grids were applied in the dilute zone and tetrahedral grids were applied in the dense region near the inlets. The total mesh number is about 51,000. After the time step-independence analysis, a time step of 1×10^{-3} s with the convergence criterion of 1×10^{-4} was set for this study.

As shown in Figure 17, the velocity inlet boundary condition was adopted at both of the steam inlet and feed inlet. The outflow boundary condition was used at the outlet. At the walls, the no-slip wall condition was assumed [22,45,48]. The maximum particle packing was limited to 0.64. Physical parameters of the gas and solid mixtures obeyed the volume/mass-weighted-mixing law. The fuel reactor was initially filled with steam (H_2O), the solid particles of coal and oxygen carrier were then fed into the reactor at the beginning of a simulation.

The oxygen carrier material used in the simulations was assumed to be the Norwegian ilmenite [22,24,25]. The particle diameter of the ilmenite was assumed to be uniform (150 μm). The fuel adopted for the simulation was a Colombian bituminous coal [22,70] with a uniform diameter of 200 μm. Further details on the properties of the Norwegian ilmenite and the Colombian coal can be found from our previous publication [22].

4. Conclusions

A comprehensive three-dimensional numerical model including the kinetic theory of granular flow and complicated gas–solid reactions was developed to simulate the iG-CLC process in a circulating fluidized bed (CFB) fuel reactor. Extending from the previous basic simulations and after further validations, the model has been used to study the effects of some important operating conditions, i.e., solids flux, steam flow and operating pressure, on the gas–solid flow behaviors, CO_2 concentration and fuel conversion. The following conclusions can be drawn from the present study:

(1) The single-loop conversion of carbon and CO_2 dry-basis concentration at the fuel reactor outlet under the reference condition are comparable to those from the previous experimental system, demonstrating the CFB riser with high solids flux is a potential candidate for the fuel reactor of iG-CLC.

(2) An increase in the solids flux results in an increase in the CO_2 concentration at the outlet, which is believed to be mainly due to the increased solids holdup and better gas–solid contacts that promote gas–solid reactions. However, a higher solids flux has a slightly negative effect on the single-loop carbon conversion. This is mainly due to the decrease in the solids residence time.

(3) A decrease in the steam flow gives rise to an increase in CO_2 concentration at the fuel reactor outlet because of the higher solids holdup and the lower gas velocity, which further enhances the gas–solid contacts and the residence time of gas for reactions. In addition, a decrease in the steam flow promotes the single-loop conversion of carbon mainly as a result of the increased solids residence time. However, in order to ensure the circulation and fluidization of the particles, there exists a minimum requirement for the steam flow under the conditions of the given operating temperature, pressure and particle flows.

(4) An increase in the operating pressure leads to an increase in the CO_2 concentration at the outlet of the fuel reactor due to the higher concentrations of gasification intermediates, higher solids holdup and the lower gas velocity, which promote the reduction reactions (8)–(10) with the oxygen carrier. The elevated steam partial pressure and longer gas residence time facilitate the char gasification. Moreover, the elevated pressure leads to a decrease in solids velocity, thus longer solids residence time and higher carbon conversion. Therefore, it should be beneficial

for a practical CFB fuel reactor of iG-CLC system to be designed and operated under a certain pressurized conditions.

Author Contributions: Conceptualization, X.W. and B.J.; investigation, X.W.; formal analysis, X.W. and H.L.; supervision, B.Z.; writing—original draft preparation, X.W.; writing—review and editing, H.L., and Y.Z.

Funding: This research was funded by the National Natural Science Foundation of China (grant numbers 51806035, 51741603, 51676038), the Natural Science Fund project in Jiangsu Province (grant number BK20170669), the Fundamental Research Funds for the Central Universities (grant number 2242018K40117), and the Guangdong Provincial Key Laboratory of New and Renewable Energy Research and Development (grant number Y707s41001).

Conflicts of Interest: The authors declare no conflict of interest.

Nomenclature

A_f	sectional area of the feed inlet [m^2]
A_s	sectional area of the fuel reactor [m^2]
f_i	dry basis concentration of gas component
$\vec{g}$	acceleration due to gravity [m/s^2]
G_p	solids flux [kg/(m^2·s)]
H	specific enthalpy [J/kg]
J_i	diffusion flux of species i [kg/(m^2·s)]
p	pressure [Pa]
P_0	operating pressure [MPa]
$Q_{C,out}$	mass flux of unreacted char at the outlet [kg/h]
$Q_{Coal,in}$	mass flux of coal at the solids inlet [kg/h]
Q_h	steam flow under the operating temperature and pressure [m^3/h]
$\dot{m}$	mass source term [kg/(m^3·s)]
S_i	net rate of production of species i [kg/(m^3·s)]
$U_{p,f}$	particle velocity at the feed inlet [m/s]
v	velocity [m/s]
x_i	molar fraction of species i in the gas phase
X_C	single-loop conversion of carbon [%]
Y	mass fraction
α	volume fraction
β	drag [kg/(m^3·s)]
λ	thermal conductivity [W/(m^2·K)]
ρ	density [kg/m^3]
$\bar{\bar{\tau}}$	stress-strain tensor [Pa]

References

1. Lyngfelt, A.; Leckner, B.; Mattisson, T. A fluidized-bed combustion process with inherent CO$_2$ separation; application of chemical-looping combustion. *Chem. Eng. Sci.* **2001**, *56*, 3101–3113. [CrossRef]
2. Abad, A.; Mattisson, T.; Lyngfelt, A.; Johansson, M. The use of iron oxide as oxygen carrier in a chemical-looping reactor. *Fuel* **2007**, *86*, 1021–1035. [CrossRef]
3. Mattisson, T.; García-Labiano, F.; Kronberger, B.; Lyngfelt, A.; Adánez, J.; Hofbauer, H. Chemical-Looping Combustion using syngas as fuel. *Int. J. Greenh. Gas Control* **2007**, *1*, 158–169. [CrossRef]
4. Ishida, M.; Jin, H.; Okamoto, T. A fundamental study of a new kind of medium material for chemical-looping combustion. *Energy Fuels* **1996**, *10*, 958–963. [CrossRef]
5. Jin, H.; Okamoto, T.; Ishida, M. Development of a novel chemical-looping combustion: Synthesis of a solid looping material of NiO/NiAl$_2$O$_4$. *Ind. Eng. Chem. Res.* **1999**, *38*, 126–132. [CrossRef]
6. Cho, P.; Mattisson, T.; Lyngfelt, A. Comparison of iron-, nickel-, copper-and manganese-based oxygen carriers for chemical-looping combustion. *Fuel* **2004**, *83*, 1215–1225. [CrossRef]
7. Garcia-Labiano, F.; Adanez, J.; de Diego, L.F.; Gayán, P.; Abad, A. Effect of pressure on the behavior of copper-, iron-, and nickel-based oxygen carriers for chemical-looping combustion. *Energy Fuels* **2006**, *20*, 26–33. [CrossRef]

8. De Diego, L.F.; García-Labiano, F.; Gayán, P.; Celaya, J.; Palacios, J.M.; Adánez, J. Operation of a 10 kW$_{th}$ chemical-looping combustor during 200 h with a CuO-Al$_2$O$_3$ oxygen carrier. *Fuel* **2007**, *86*, 1036–1045. [CrossRef]
9. Adánez, J.; Dueso, C.; de Diego, L.F.; García-Labiano, F.; Gayán, P.; Abad, A. Methane combustion in a 500 W$_{th}$ chemical-looping combustion system using an impregnated Ni-based oxygen carrier. *Energy Fuels* **2008**, *23*, 130–142. [CrossRef]
10. Kolbitsch, P.; Bolhàr-Nordenkampf, J.; Pröll, T.; Hofbauer, H. Operating experience with chemical looping combustion in a 120kW dual circulating fluidized bed (DCFB) unit. *Int. J. Greenh. Gas Control* **2010**, *4*, 180–185. [CrossRef]
11. Ma, J.; Zhao, H.; Tian, X.; Wei, Y.; Zhang, Y.; Zheng, C. Continuous Operation of Interconnected Fluidized Bed Reactor for Chemical Looping Combustion of CH$_4$ Using Hematite as Oxygen Carrier. *Energy Fuels* **2015**, *29*, 3257–3267. [CrossRef]
12. Berguerand, N.; Lyngfelt, A. Design and operation of a 10 kW$_{th}$ chemical-looping combustor for solid fuels-testing with South African coal. *Fuel* **2008**, *87*, 2713–2726. [CrossRef]
13. Leion, H.; Mattisson, T.; Lyngfelt, A. Solid fuels in chemical-looping combustion. *Int. J. Greenh. Gas Control* **2008**, *2*, 180–193. [CrossRef]
14. Shen, L.H.; Wu, J.H.; Xiao, J. Experiments on chemical looping combustion of coal with a NiO based oxygen carrier. *Combust. Flame* **2009**, *156*, 721–728. [CrossRef]
15. Fan, L.S.; Li, F. Chemical looping technology and its fossil energy conversion applications. *Ind. Eng. Chem. Res.* **2010**, *49*, 10200–10211. [CrossRef]
16. Abad, A.; Gayán, P.; de Diego, L.F.; García-Labiano, F.; Adánez, J. Fuel reactor modelling in chemical-looping combustion of coal: 1. Model formulation. *Chem. Eng. Sci.* **2013**, *87*, 277–293. [CrossRef]
17. García-Labiano, F.; de Diego, L.F.; Gayán, P.; Abad, A.; Adánez, J. Fuel reactor modelling in chemical-looping combustion of coal: 2-simulation and optimization. *Chem. Eng. Sci.* **2013**, *87*, 173–182. [CrossRef]
18. Forero, C.R.; Gayán, P.; de Diego, L.F.; Abad, A.; García-Labiano, F.; Adánez, J. Syngas combustion in a 500 Wth chemical-looping combustion system using an impregnated Cu-based oxygen carrier. *Fuel Process. Technol.* **2009**, *90*, 1471–1479. [CrossRef]
19. Jin, H.; Ishida, M. A new type of coal gas fueled chemical-looping combustion. *Fuel* **2004**, *83*, 2411–2417. [CrossRef]
20. Spallina, V.; Gallucci, F.; Romano, M.C.; Chiesa, P.; Lozza, G.; van Sint Annaland, M. Investigation of heat management for CLC of syngas in packed bed reactors. *Chem. Eng. J.* **2013**, *225*, 174–191. [CrossRef]
21. Hamers, H.P.; Romano, M.C.; Spallina, V.; Chiesa, P.; Gallucci, F.; van Sint Annaland, M. Energy analysis of two stage packed-bed chemical looping combustion configurations for integrated gasification combined cycles. *Energy* **2015**, *85*, 489–502. [CrossRef]
22. Wang, X.; Jin, B.; Zhang, Y.; Zhang, Y.; Liu, X. Three Dimensional Modeling of a Coal-Fired Chemical Looping Combustion Process in the Circulating Fluidized Bed Fuel Reactor. *Energy Fuels* **2013**, *27*, 2173–2184. [CrossRef]
23. Wang, X.; Jin, B.; Liu, X.; Zhang, Y.; Liu, H. Experimental investigation on flow behaviors in a novel in situ gasification chemical looping combustion apparatus. *Ind. Eng. Chem. Res.* **2013**, *52*, 14208–14218. [CrossRef]
24. Abad, A.; Adánez, J.; Cuadrat, A.; García-Labiano, F.; Gayán, P.; Luis, F. Kinetics of redox reactions of ilmenite for chemical-looping combustion. *Chem. Eng. Sci.* **2011**, *66*, 689–702. [CrossRef]
25. Adánez, J.; Cuadrat, A.; Abad, A.; Gayán, P.; de Diego, L.; García-Labiano, F. Ilmenite activation during consecutive redox cycles in chemical-looping combustion. *Energy Fuels* **2010**, *24*, 1402–1413. [CrossRef]
26. Xiao, R.; Song, Q.; Song, M.; Lu, Z.; Zhang, S.; Shen, L. Pressurized chemical-looping combustion of coal with an iron ore-based oxygen carrier. *Combust. Flame* **2010**, *157*, 1140–1153. [CrossRef]
27. Cuadrat, A.; Abad, A.; Adánez, J.; de Diego, L.; García-Labiano, F.; Gayán, P. Behavior of ilmenite as oxygen carrier in chemical-looping combustion. *Fuel Process. Technol.* **2012**, *94*, 101–112. [CrossRef]
28. Cuadrat, A.; Abad, A.; García-Labiano, F.; Gayán, P.; de Diego, L.; Adánez, J. Effect of operating conditions in Chemical-Looping Combustion of coal in a 500 W$_{th}$ unit. *Int. J. Greenh. Gas Control* **2012**, *6*, 153–163. [CrossRef]
29. Bayham, S.; McGiveron, O.; Tong, A.; Chung, E.; Kathe, M.; Wang, D.; Zeng, L.; Fan, L.S. Parametric and dynamic studies of an iron-based 25-kW$_{th}$ coal direct chemical looping unit using sub-bituminous coal. *Appl. Energy* **2015**, *145*, 354–363. [CrossRef]

30. Linderholm, C.; Schmitz, M. Chemical-looping combustion of solid fuels in a 100 kW dual circulating fluidized bed system using iron ore as oxygen carrier. *J. Environ. Chem. Eng.* **2016**, *4*, 1029–1039. [CrossRef]
31. Ma, J.; Zhao, H.; Tian, X.; Wei, Y.; Rajendran, S.; Zhang, Y.; Bhattacharya, S.; Zheng, C. Chemical looping combustion of coal in a 5 kW$_{th}$ interconnected fluidized bed reactor using hematite as oxygen carrier. *Appl. Energy* **2015**, *157*, 304–313. [CrossRef]
32. Pérez-Vega, R.; Abad, A.; García-Labiano, F.; Gayán, P.; Luis, F.; Adánez, J. Coal combustion in a 50 kW$_{th}$ Chemical Looping Combustion unit: Seeking operating conditions to maximize CO_2 capture and combustion efficiency. *Int. J. Greenh. Gas Control* **2016**, *50*, 80–92. [CrossRef]
33. Ströhle, J.; Orth, M.; Epple, B. Design and operation of a 1 MW$_{th}$ chemical looping plant. *Appl. Energy* **2014**, *113*, 1490–1495. [CrossRef]
34. Thon, A.; Kramp, M.; Hartge, E.-U.; Heinrich, S.; Werther, J. Operational experience with a system of coupled fluidized beds for chemical looping combustion of solid fuels using ilmenite as oxygen carrier. *Appl. Energy* **2014**, *118*, 309–317. [CrossRef]
35. Xiao, R.; Chen, L.; Saha, C.; Zhang, S.; Bhattacharya, S. Pressurized chemical-looping combustion of coal using an iron ore as oxygen carrier in a pilot-scale unit. *Int. J. Greenh. Gas Control* **2012**, *10*, 363–373. [CrossRef]
36. Wang, X.; Jin, B.; Zhu, X.; Liu, H. Experimental Evaluation of a Novel 20 kW$_{th}$ in Situ Gasification Chemical Looping Combustion Unit with an Iron Ore as the Oxygen Carrier. *Ind. Eng. Chem. Res.* **2016**, *55*, 11775–11784. [CrossRef]
37. Gayán, P.; Abad, A.; de Diego, L.F.; García-Labiano, F.; Adánez, J. Assessment of technological solutions for improving chemical looping combustion of solid fuels with CO_2 capture. *Chem. Eng. J.* **2013**, *233*, 56–69. [CrossRef]
38. Abad, A.; Mendiara, T.; Gayán, P.; García-Labiano, F.; de Diego, L.F.; Bueno, J.A.; Pérez-Vega, R.; Adánez, J. Comparative Evaluation of the Performance of Coal Combustion in 0.5 and 50 kW$_{th}$ Chemical Looping Combustion units with ilmenite, redmud or iron ore as oxygen carrier. *Energy Procedia* **2017**, *114*, 285–301. [CrossRef]
39. Adánez, J.; Abad, A.; Mendiara, T.; Gayán, P.; de Diego, L.F.; García-Labiano, F. Chemical looping combustion of solid fuels. *Prog. Energy Combust.* **2018**, *65*, 6–66. [CrossRef]
40. Cheng, M.; Sun, H.; Li, Z.; Cai, N. Annular carbon stripper for chemical-looping combustion of coal. *Ind. Eng. Chem. Res.* **2017**, *56*, 1580–1593. [CrossRef]
41. Wang, P.; Means, N.; Howard, B.H.; Shekhawat, D.; Berry, D. The reactivity of CuO oxygen carrier and coal in Chemical-Looping with Oxygen Uncoupled (CLOU) and In-situ Gasification Chemical-Looping Combustion (iG-CLC). *Fuel* **2018**, *217*, 642–649. [CrossRef]
42. Chen, X.; Shi, D.; Gao, X.; Luo, Z. A fundamental CFD study of the gas–solid flow field in fluidized bed polymerization reactors. *Powder Technol.* **2011**, *205*, 276–288. [CrossRef]
43. Maghrebi, R.; Yaghobi, N.; Seyednejadian, S.; Tabatabaei, M.H. CFD modeling of catalyst pellet for oxidative coupling of methane: Heat transfer and reaction. *Particuology* **2013**, *11*, 506–513. [CrossRef]
44. Shah, M.T.; Utikar, R.P.; Pareek, V.K. CFD study: Effect of pulsating flow on gas–solid hydrodynamics in FCC riser. *Particuology* **2017**, *31*, 25–34. [CrossRef]
45. Wang, X.; Jin, B.; Wang, Y.; Hu, C. Three-dimensional multi-phase simulation of the mixing and segregation of binary particle mixtures in a two-jet spout fluidized bed. *Particuology* **2015**, *22*, 185–193. [CrossRef]
46. Wang, X.; Jin, B.; Zhong, W. Three-dimensional simulation of fluidized bed coal gasification. *Chem. Eng. Process.* **2009**, *48*, 695–705. [CrossRef]
47. Zhou, W.; Zhao, C.S.; Duan, L.B.; Qu, C.R.; Chen, X.P. Two-dimensional computational fluid dynamics simulation of coal combustion in a circulating fluidized bed combustor. *Chem. Eng. J.* **2011**, *166*, 306–314. [CrossRef]
48. Deng, Z.; Xiao, R.; Jin, B.; Song, Q.; Huang, H. Multiphase CFD Modeling for a Chemical Looping Combustion Process (Fuel Reactor). *Chem. Eng. Technol.* **2008**, *31*, 1754–1766. [CrossRef]
49. Jung, J.; Gamwo, I. Multiphase CFD-based models for chemical looping combustion process: Fuel reactor modeling. *Powder Technol.* **2008**, *183*, 401–409. [CrossRef]
50. Mahalatkar, K.; Kuhlman, J.; Huckaby, E.; O'Brien, T. Computational fluid dynamic simulations of chemical looping fuel reactors utilizing gaseous fuels. *Chem. Eng. Sci.* **2011**, *66*, 469–479. [CrossRef]

51. Wang, X.; Jin, B.; Zhang, Y.; Zhong, W.; Yin, S. Multiphase Computational Fluid Dynamics (CFD) Modeling of Chemical Looping Combustion Using a CuO/Al$_2$O$_3$ Oxygen Carrier: Effect of Operating Conditions on Coal Gas Combustion. *Energy Fuels* **2011**, *25*, 3815–3824. [CrossRef]
52. Wang, X.; Jin, B.; Zhong, W.; Zhang, Y.; Song, M. Three-dimensional simulation of a coal gas fueled chemical looping combustion process. *Int. J. Greenh. Gas Control* **2011**, *5*, 1498–1506. [CrossRef]
53. Wang, S.; Lu, H.; Zhao, F.; Liu, G. CFD studies of dual circulating fluidized bed reactors for chemical looping combustion processes. *Chem. Eng. J.* **2014**, *236*, 121–130. [CrossRef]
54. Mahalatkar, K.; Kuhlman, J.; Huckaby, E.; O'Brien, T. CFD simulation of a chemical-looping fuel reactor utilizing solid fuel. *Chem. Eng. Sci.* **2011**, *66*, 3617–3627. [CrossRef]
55. Su, M.; Zhao, H.; Ma, J. Computational fluid dynamics simulation for chemical looping combustion of coal in a dual circulation fluidized bed. *Energy Convers. Manag.* **2015**, *105*, 1–12. [CrossRef]
56. Shao, Y.; Zhang, Y.; Wang, X.; Wang, X.; Jin, B.; Liu, H. Three-dimensional full loop modeling and optimization of an in situ gasification chemical looping combustion system. *Energy Fuels* **2017**, *31*, 13859–13870. [CrossRef]
57. Alobaid, F.; Ohlemüller, P.; Ströhle, J.; Epple, B. Extended Euler–Euler model for the simulation of a 1 MWth chemical–looping pilot plant. *Energy* **2015**, *93*, 2395–2405. [CrossRef]
58. May, J.; Alobaid, F.; Ohlemüller, P.; Stroh, A.; Ströhle, J.; Epple, B. Reactive two–fluid model for chemical–looping combustion–Simulation of fuel and air reactors. *Int. J. Greenh. Gas Control* **2018**, *76*, 175–192. [CrossRef]
59. Pärssinen, J.; Zhu, J. Particle velocity and flow development in a long and high-flux circulating fluidized bed riser. *Chem. Eng. Sci.* **2001**, *56*, 5295–5303. [CrossRef]
60. Jin, B.; Wang, X.; Zhong, W.; Tao, H.; Ren, B.; Xiao, R. Modeling on high-flux circulating fluidized bed with Geldart group B particles by kinetic theory of granular flow. *Energy Fuels* **2010**, *24*, 3159–3172. [CrossRef]
61. Wang, X.; Jin, B.; Zhong, W.; Zhang, M.; Huang, Y.; Duan, F. Flow behaviors in a high-flux circulating fluidized bed. *Int. J. Chem. React. Eng.* **2008**, *6*, A79. [CrossRef]
62. Wang, X.; Jin, B.; Liu, H.; Wang, W.; Liu, X.; Zhang, Y. Optimization of in Situ Gasification Chemical Looping Combustion through Experimental Investigations with a Cold Experimental System. *Ind. Eng. Chem. Res.* **2015**, *54*, 5749–5758. [CrossRef]
63. Contractor, R.M.; Patience, G.S.; Garnett, D.I.; Horowitz, H.S.; Sisler, G.M.; Bergna, H.E. A new process for n-butane oxidation to maleic anhydride using a circulating fluidized bed reactor. In *Circulating Fluidized Bed Technology IV*; Avidan, A., Ed.; AIChE: New York, NY, USA, 1994.
64. Mei, J.S.; Shadle, L.J.; Yue, P.; Monazam, E.R. Hydrodynamics of a transport reactor operating in dense suspension upflow conditions for coal combustion applications. In Proceedings of the 18th International Conference on Fluidized Bed Combustion, Toronto, ON, Canada, 22–25 May 2005.
65. Morton, F.; Pinkston, T.; Salazar, N.; Stalls, D. Orlando gasification project: Demonstration of a nominal 285 MW coal-based transport gasifier. In Proceedings of the 23rd Annual International Pittsburgh Coal Conference, Pittsburgh, PA, USA, 25–28 September 2006.
66. Li, Z.; Wu, C.; Wei, F.; Jin, Y. Experimental study of high-density gas—Solids flow in a new coupled circulating fluidized bed. *Powder Technol.* **2004**, *139*, 214–220. [CrossRef]
67. Malcus, S.; Cruz, E.; Rowe, C.; Pugsley, T. Radial solid mass flux profiles in a high-suspension density circulating fluidized bed. *Powder Technol.* **2002**, *125*, 5–9. [CrossRef]
68. Liu, G.; Niksa, S. Coal conversion submodels for design applications at elevated pressures. Part II. Char gasification. *Prog. Energy Combust. Sci.* **2004**, *30*, 679–717. [CrossRef]
69. Yin, S.; Jin, B.; Zhong, W.; Lu, Y.; Zhang, Y.; Shao, Y.; Liu, H. Solids holdup of high flux circulating fluidized bed at elevated pressure. *Chem. Eng. Technol.* **2012**, *35*, 904–910. [CrossRef]
70. Cuadrat, A.; Abad, A.; Gayán, P.; de Diego, L.F.; García-Labiano, F.; Adánez, J. Theoretical approach on the CLC performance with solid fuels: Optimizing the solids inventory. *Fuel* **2012**, *97*, 536–551. [CrossRef]
71. Benyahia, S.; Syamlal, M.; O'Brien, T.J. Study of the ability of multiphase continuum models to predict core-annulus flow. *AIChE J.* **2007**, *53*, 2549–2568. [CrossRef]
72. Gidaspow, D.; Bezburuah, R.; Ding, J. Hydrodynamics of circulating fluidized beds: Kinetic theory approach. In Proceedings of the 7th Engineering Foundation Conference on Fluidization, Brisbane, Australia, 3–8 May 1992.

73. Lun, C.K.K.; Savage, S.B.; Jeffrey, D.J.; Chepurniy, N. Kinetic theories for granular flow: Inelastic particles in Couette flow and slightly inelastic particles in a general flowfield. *J. Fluid Mech.* **1984**, *140*, 223–256. [CrossRef]
74. Donskoi, E.; McElwain, D.; Wibberley, L. Estimation and modeling of parameters for direct reduction in iron ore/coal composites: Part II. Kinetic parameters. *Metall. Mater. Trans. B* **2003**, *34*, 255–266. [CrossRef]
75. Smoot, L.; Smith, P. *Coal Combustion and Gasification*; Plenum Press: New York, NY, USA, 1985.
76. Carrero-Mantilla, J.; Llano-Restrepo, M. Chemical equilibria of multiple-reaction systems from reaction ensemble Monte Carlo simulation and a predictive equation of state: Combined hydrogenation of ethylene and propylene. *Fluid Phase Equilibr.* **2006**, *242*, 189–203. [CrossRef]

© 2018 by the authors. Licensee MDPI, Basel, Switzerland. This article is an open access article distributed under the terms and conditions of the Creative Commons Attribution (CC BY) license (http://creativecommons.org/licenses/by/4.0/).

Article

A Novel Method for the Prediction of Erosion Evolution Process Based on Dynamic Mesh and Its Applications

Yunshan Dong [1], Zongliang Qiao [1], Fengqi Si [1,*], Bo Zhang [1,*], Cong Yu [1] and Xiaoming Jiang [2]

1. Key Laboratory of Energy Thermal Conversion and Control of Ministry of Education, School of Energy and Environment, Southeast University, Nanjing 210096, China; ysh_dong@126.com (Y.D.); qiaozongliang@seu.edu.cn (Z.Q.); congy@seu.edu.cn (C.Y.)
2. Datang Nanjing Environmental Protection Technology Co., Ltd., Nanjing 211100, China; jiangxm@dteg.com.cn
* Correspondence: fqsi@seu.edu.cn (F.S.); bozhang@seu.edu.cn (B.Z.)

Received: 29 August 2018; Accepted: 28 September 2018; Published: 30 September 2018

Abstract: Particle erosion is a commonly occurring phenomenon, and it plays a significantly important role in service life. However, few simulations have replicated erosion, especially the detailed evolution process. To address this complex issue, a new method for establishing the solution of the erosion evolution process was developed. The approach is introduced with the erosion model and the dynamic mesh. The erosion model was applied to estimate the material removal of erosion, and the dynamic mesh technology was used to demonstrate the surface profile of erosion. Then, this method was applied to solve a typical case—the erosion surface deformation and the expiry period of an economizer bank in coal-fired power plants. The mathematical models were set up, including gas motion, particle motion, particle-wall collision, and erosion. Such models were solved by computational fluid dynamics (CFD) software (ANSYS FLUENT), which describes the evolution process of erosion based on the dynamic mesh. The results indicate that: (1) the prediction of the erosion profile calculated by the dynamic mesh is in good agreement with that on-site; (2) the global/local erosion loss and the maximum erosion depth is linearly related to the working time at the earlier stage, but the growth of the maximum erosion depth slows down gradually in the later stage; (3) the reason for slowing down is that the collision point trajectory moves along the increasing direction of the absolute value of θ as time increases; and (4) the expiry period is shortened as the ash diameter increases.

Keywords: erosion evolution; erosion rate; dynamic mesh; CFD; economizer; expiry period

---

1. Introduction

Erosion is mechanical damage resulting from the impact of particles carried by fluid [1]. When a particle with a certain speed strikes a solid surface, the impact region of the surface will be deformed [2–4]. Subsequently, the surface deformation translates into material removal, which shortens the service life, such as the SCR (selective catalytic reduction) catalyst and the economizer in Figure 1, among others. Erosion is a complex problem caused by numerous factors, such as particle velocity, impact angle, particle size, and particle shape.

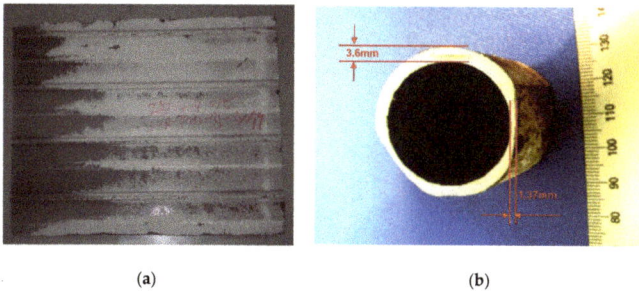

Figure 1. Particle erosion. (**a**) SCR catalyst; (**b**) economizer (reprinted with permission from reference [5]. Copyright 2018 ELSEVIER).

Considerable studies have been performed on erosion. Finnie [6,7] and Bitter et al. [8,9] presented the earliest work focusing on theoretical erosion mechanism. Finnie's erosion model was based on the experimental phenomenon of erosion and the models of the micro-cutting mechanism. However, at the impact angle of 90°, it predicted no erosion rate at all, and the analyzed data clearly indicated that it is incorrect. Bitter's erosion model was proposed with respect to deformation and cutting wear, despite lacking the support of any physical models. Huang et al. [10] addressed these shortages, deriving a phenomenological erosion model by analyzing the normal and tangential forces acting on the abrasive particle. Huang's erosion model has been increasingly acknowledged by researchers. In addition to the theoretical erosion model, numerous erosion equations have been developed based on experimental tests. Most of these erosion equations have been presented regarding the function of the velocity exponent and the impact angle. Grant and Tabakoff [11] developed an empirical erosion equation. This model pointed out that the coefficient of restitution should be incorporated into the equation, owing to multiple times of impingement by the particle. Okal et al. [12,13] conducted erosion tests for a wide range of materials and erodent particles, considering both material hardness and the load relaxation ratio. Many parameters were involved in this model, although the parameters are hard to acquire. Besides these, Haugen [14], Ahlert [15], Mclaury [16], and Zhang [17] at E/CRC of the University of Tulsa also proposed different empirical erosions, and Shamshirband et al. [18] adapted the ANFIS (adaptive neuro-fuzzy inference system, a type of neural network) model to precisely predict the total and maximum erosion rate.

The aforementioned work focused on the erosion ratio. However, the evolution process of erosion has been little studied in the research, mainly because no methods or models have described long periods of erosion or shape evolution, except for the experiments. However, such experiments take too much time and material power. If feasible methods or models were available to describe the evolution process, the research would make progress.

In this study, the material removal of erosion was investigated by CFD (computational fluid dynamics software) coupled with the UDF (user defined function) of the particle motion and erosion model. It is worth highlighting that the detailed evolution process of erosion will be realized by the dynamic mesh technique. This DPM (discrete phase model) approach, coupling the erosion model and dynamic mesh, combines the material removal and the mesh deformation quantitatively, which can demonstrate the node displacement of the mesh in the erosion region and the new flow field. Accordingly, the flow field and the particle trajectory are transformed at all times. Based on those concepts, the evolution of the erosion loss and erosion depth was investigated by the CFD-DPM approach firstly. The erosion profile was achieved secondly.

Moreover, this paper applies the method to discuss the expiry period of the economizer in modern coal-fired power plants. The economizer is used to improve the overall thermal efficiency in the gas-water exchangers, exacting residual heat energy from the flue gas and transferring the energy to the feed water [19]. However, the flue gas includes ash particles, corrosive gas, temperature oscillation, and so on. Those factors respectively contribute to ash erosion, corrosion fatigue, thermal fatigue, and so forth, which may ultimately lead to bursting of economizer tubes [20]. Among these, ash erosion plays a significantly important role in the expiry period of the economizer. Therefore, this work will benefit the design of the economizer and guide the boiler operation.

2. Method Description

Figure 2 demonstrates the calculative process of erosion evolution. The process includes the flow calculation, particle tracking, material removal calculation coupled with the UDF of the erosion model, erosion profile calculation with the UDF of the dynamic mesh, and the ALE (arbitrary *Lagrange–Euler*) calculation. ANSYS FLUENT calculates the trajectory of the discrete particle by integrating the force balance on the particle, which is done in the *Lagrangian* frame. Then, the collision point of the particle-wall collision is tracked, and the particle-wall collision is translated into the material removal in the UDF of the erosion model. Then, the material removal is converted to wall-mesh deformation in the UDF of the dynamic mesh. Dynamic mesh is universally applied for the fluid–structure interaction [21], flapping flight [22] and so forth. However, this dynamic mesh technology is rarely introduced to the erosion profile that is formed by wall-mesh deformation. Figure 3 shows the mesh deformation before and after the particle-wall collision by the dynamic mesh. The value of the material removal is stored in the center of the boundary mesh. However, the dynamic mesh moves the node of the boundary mesh, and the value in the node is interpolated by the value in the center. In this interpolation, this paper keeps the elementary volume constant, as follows:

$$h_{i,j} = \frac{1/4\Delta t \sum_{k=0}^{1} \sum_{l=0}^{1} Q_{i-1/2+k,j-1/2+l}^{n}}{1/4\rho_m \sum_{k=0}^{1} \sum_{l=0}^{1} A_{i-1/2+k,j-1/2+l}^{n-1}} \quad (1)$$

$Q_{i-1/2,j-1/2}$ is the rate of the material removal, the function of the particle mass flow rate m_p and the erosion rate E_r in the $[i-1/2, j-1/2]$ element mesh; $A_{i-1/2,j-1/2}$ is the area of the $[i-1/2, j-1/2]$ element mesh; n is the time step; Δt is the time step size.

The ALE calculation is used to correct the flux after the mesh deformation. In the ALE description, the nodes of the computational mesh may be moved with the continuum in a normal *Lagrangian* fashion, or be held fixed in a *Eulerian* manner [23,24]. At last, this paper adopts the unsteady calculation, and the erosion effect of the time step amounts to 1day. It is possible to regard the erosion rate E_r as the constant during 1day. Additionally, when computation time t is at the maximum time t_{max} (2 years in this paper) given, the calculation is terminated.

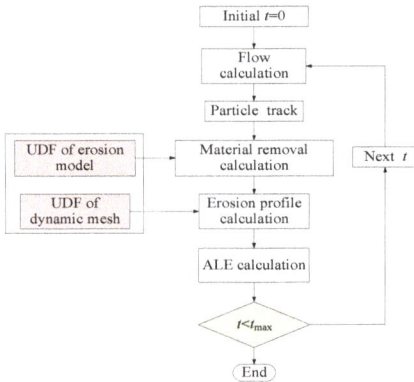

Figure 2. Calculative process of erosion evolution.

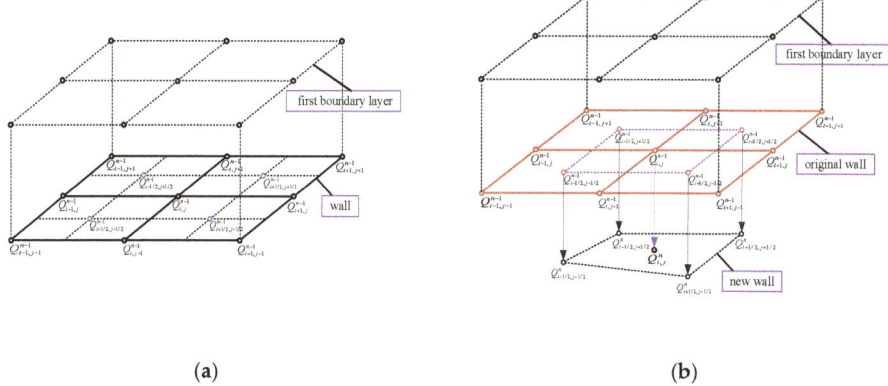

Figure 3. Erosion profile calculation with dynamic mesh: (**a**) before collision; (**b**) after collision.

When the erosion profile is transformed, the flow field, especially the boundary layer, will be changed also. The specific mathematical relation between the material removal and the wall-mesh deformation is in Formula (1). The wall-mesh deformation leads conversely to the modification of the particle trajectory. This is the coupled calculation of multi-physics, including the flow, particle motion, and erosion.

3. Application

3.1. Mathematical Model

3.1.1. Flow Configuration

The economizer of a 600MW coal-fired power plant was studied in this work. It is composed of 135 rows of S-shaped circular tubes and made from SA 210 GRA1(N); the other detailed parameters are shown in Table 1. Considering the periodic distribution of the tube bank in Figure 4, this paper takes the shaded region as the research unit.

Figure 5 gives the three-dimensional diagram of the research unit and the boundary conditions. Boundary conditions were set as follows: velocity-inlet, pressure-outlet, and wall. As mentioned earlier, there exists periodicity in the Y direction. Meanwhile, along the tube (Z direction), the flow is similar. Thus, the periodic boundary is loaded onto the Y direction and the Z direction.

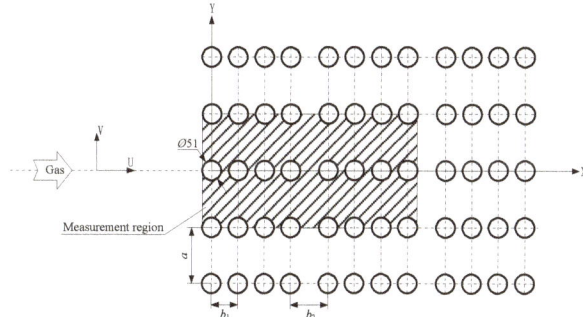

Figure 4. Schematic diagram of the economizer tube bank.

Table 1. Structural parameters of the economizer tube [25].

Material	Rows	Diameter/mm	Transversal Pitch a/mm	Longitudinal Pitch b_1/mm	Longitudinal Pitch b_2/mm
SA 210 GrA1(N)	135	Φ51×6	144	102	69

The velocity is selected at 8.12 m/s, which represents 100% of the boiler THA (turbine heat acceptance) load. The diameters of the particles are selected at 50 μm, 100 μm, and 250 μm, with the same mass flow. The mass flow of ash is set to 1.52×10^{-2} kg/s, equal to 32 g/m³ of the gas.

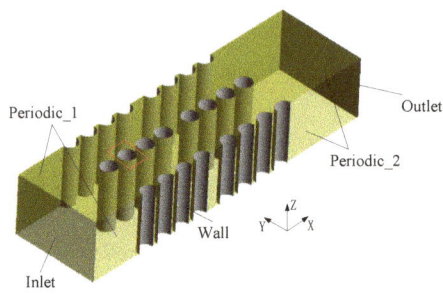

Figure 5. Three-dimensional diagram of the economizer unit.

Considering the effect of the boundary layer on the flow computation, the grid of the tube wall is refined in Figure 6 (enlarged from the red frame of Figure 5). Then it will capture the internal flow field information on the boundary layer. Moreover, the grid-independency study gives the appropriate grid number, whose mean error is controlled at less than 2%. The grid number ensures the accuracy and rapidity of the calculation.

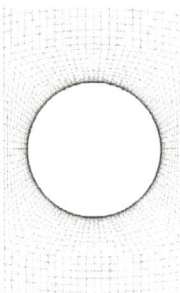

Figure 6. Grid refinement of the tube wall.

3.1.2. Governing Equations

Gas Motion Model

Gas from the furnace flows around the tube bank of the economizer. The flow field changes, but it meets *Navier–Stokes* equations. Considering the flow as a turbulent flow, the two-equation turbulent model (standard *k-ε* model) is introduced. The *Navier–Stokes* equations coupled with the standard *k-ε* turbulence model is employed to solve the gas motion. *Navier–Stokes* equations of gas are as follows [26,27]:

$$\frac{\partial(\rho_f u_i)}{\partial x_i} = 0 \qquad (2)$$

$$\frac{\partial(\rho_f u_i u_j)}{\partial x_j} = \rho_f g - \frac{\partial p}{\partial x_i} + \frac{\partial \tau_{ij}}{\partial x_j} \tag{3}$$

$$\frac{\partial(\rho_f e u_i)}{\partial x_i} = -\frac{\partial(u_i p)}{\partial x_i} + \frac{\partial}{\partial x_i}\left(k\frac{\partial T}{\partial x_i}\right) + \frac{\partial(u\tau_{ij})}{\partial x_i} \tag{4}$$

τ_{ij} is the stress tensor, excluding the surface pressure. The stress tensor is calculated as:

$$\tau_{ij} = \mu\left(\frac{\partial u_i}{\partial x_j} + \frac{\partial u_j}{\partial x_i}\right) - \frac{2}{3}\mu\frac{\partial u_k}{\partial x_k}\delta_{ij} \tag{5}$$

e is the total energy, consisting of the internal and kinetic energy. The total energy is calculated as:

$$e = C_p T + \frac{1}{2}V_f^2 \tag{6}$$

The standard k-ε turbulence model is as follows [28]:

$$\begin{aligned}\frac{\partial(\rho k_f u_i)}{\partial x_i} &= \frac{\partial}{\partial x_j}\left(\left(\mu + \frac{\mu_t}{\sigma_k}\right)\frac{\partial k}{\partial x_j}\right) + G_k + G_b - \rho\varepsilon \\ &\quad - Y_M + S_k\end{aligned} \tag{7}$$

$$\begin{aligned}\frac{\partial(\rho\varepsilon u_i)}{\partial x_i} &= \frac{\partial}{\partial x_j}\left(\left(\mu + \frac{\mu_t}{\sigma_\varepsilon}\right)\frac{\partial \varepsilon}{\partial x_j}\right) \\ &\quad + C_{1\varepsilon}\frac{\varepsilon}{k_f}(G_k + C_{3\varepsilon}G_b) - C_{2\varepsilon}\rho\frac{\varepsilon^2}{k} + S_\varepsilon\end{aligned} \tag{8}$$

where:

$$\mu_t = \rho C_\mu \frac{k_f^2}{\varepsilon} \tag{9}$$

Lagrangian Formulation for Particle Motion Model

The particle motion is traced by Newton's Second Law in Formula (10), driven by the drag force F_D, thermophoretic force F_T, gravity F_G, and *Saffman* lift force F_{SL} [29].

$$\frac{du_p}{dt} = F_D + F_T + F_G + F_{SL} \tag{10}$$

where,

$$F_D = \frac{18\mu}{\rho_p d_p^2}\frac{C_D Re}{24}(u_f - u_p) \tag{11}$$

$$F_T = -\frac{6\pi d_p \mu^2 C_S\left(k_f/k_p + C_t Kn\right)}{\rho(1 + 3C_m Kn)\left(1 + 2k_f/k_p + 2C_t Kn\right)}\frac{1}{m_p T}\nabla T \tag{12}$$

$$F_G = \frac{g(\rho_p - \rho_f)}{\rho_p} \tag{13}$$

$$F_{SL} = \frac{2Kv^{1/2}\rho d_{ij}}{\rho_p d_p (d_{lk}d_{kl})^{1/4}}(u_f - u_p) \tag{14}$$

Particle-Wall Collision Model

In particle motion, the particle–particle collision and the particle-wall collision change the motion state, including velocity and direction of motion. In consideration of the economizer bank in the tail shaft flue and low concentration of ash, there are relatively few possibilities of particle–particle collision. This research mainly focuses on the particle-wall collision, neglecting the particle–particle collisions. Therefore, this study adopted the DPM approach instead of the DEM (discrete element method) approach. In the particle-wall collision, the normal and tangential components of reflected velocity are achieved by the coefficient of velocity restitution. The coefficient of velocity restitution is caused by kinetic energy loss, a function of the incidence angle α in the particle-wall collision model, as follows [30]:

$$e_n = 0.993 - 1.76\alpha + 1.56\alpha^2 - 0.49\alpha^3 \tag{15}$$

$$e_t = 0.998 - 1.66\alpha + 2.11\alpha^2 - 0.76\alpha^3 \tag{16}$$

Erosion Model

Particle-wall collision not only changes the motion state, but also removes the material. Huang's erosion model was adopted to estimate the abrasion loss from the following Formula (17). Huang's erosion model includes mechanisms of deformation damage removal and cutting removal [10]. This model also demonstrates the effect of particle size on erosion. The erosion rate E_r is the ratio of abrasion loss and particle mass.

$$E_r = C\rho_p^{0.15}(V_p \sin\alpha)^{2.3} + Dm_p^{0.1875} d_p^{-0.0625} V_p^{2.375}(\cos\alpha)^2(\sin\alpha)^{0.375} \tag{17}$$

Numerical Procedures

The spatial discretization applies the second-order upwind scheme to discretize the whole convective terms. In addition, the numerical calculation adopts the *Eulerian–Lagrangian* frame, and the SIMPLEC algorithm (a semi-implicit algorithm) is applied to pressure/velocity coupling in the *Eulerian* frame. This paper is divided into two numerical procedures, as follows:

1. Step 1: Verification of erosion model on SA210 GrA1(N);
2. Step 2: Erosion calculation of economizer unit.

3.2. Verification of Erosion Model

C and D are related to the properties of the particle and economizer material in Formula (17). C and D are determined by fitting the curve. By comparing with the erosion experimental data by the ASTM standard-tested bed in the India National Institute of Technology [31,32], the accuracy of Huang's erosion model was verified and found to be appropriate for SA 210GrA1(N) in Figures 7 and 8. Figure 7 gives the relation between the incidence angle and the erosion rate E_r. Compared to Finnie's erosion model, Huang's erosion model can not only estimate the effect of the small impact angle, but also the large angle. Figure 8 shows the relation between the particle diameter d_p and the erosion rate E_r. From Figures 7 and 8, the predicted results obtained are consistent and in good agreement with experimental results reported elsewhere. For the erosion curve of the impact angle or the ash size, the mean relative error is less than 5%, which satisfies the research requirements.

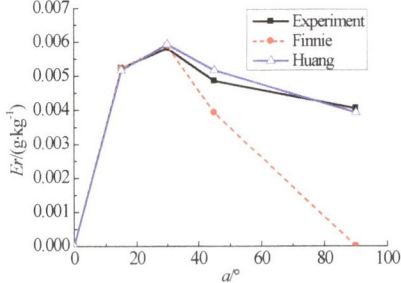

Figure 7. Erosion curve of impact angle on SA 210GrA1.

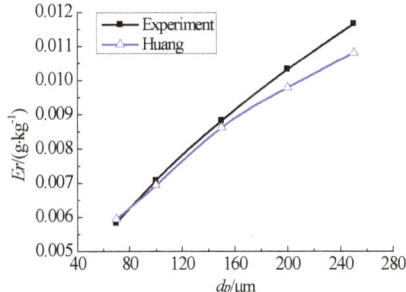

Figure 8. Erosion curve of ash size on SA 210GrA1.

4. Results and Discussion

4.1. Erosion Loss

4.1.1. Global Erosion Loss

Figure 9 shows the global erosion loss under different ash sizes. From Figure 9, the global erosion loss is linearly related to the working time. Meanwhile, as the ash size increases, the growth amplification of global erosion loss decreases ($\Delta 1 > \Delta 2$). This coincides with the fact that it is hard to change the motion of the larger particle, which has the larger inertia. In other words, the motion of the larger particle becomes smaller, and the collision point of a single particle from the injection changes little. Thus, the mass flow of ash impacting the economizer increases slowly and becomes stable gradually, which contributes to the slow growth amplification. In addition, the particle rebound or collision also impacts the global erosion loss, illustrated in Section 4.1.2.

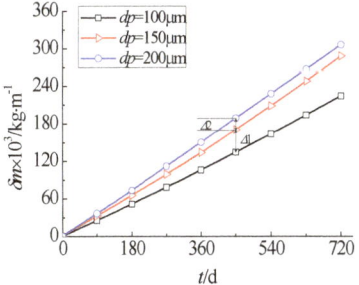

Figure 9. Global erosion loss under different ash sizes.

4.1.2. Local Erosion Loss

Figure 10 shows the local erosion loss of different rows under various ash sizes. From this figure, the local erosion loss is consistent with the global erosion loss. The local erosion loss is approximately linear to the working time. The erosion loss of the first row occupies more than half of the global erosion loss. Ash impacts the first rows, and the kinetic energy decreases. Then most of the ash particles follow the gas flow, which hinders the particle collision in the other rows.

Similarly, as the ash size increases, the growth amplification of local erosion loss decreases in the first row. Meanwhile, with the ash size increased, the mass flow of ash impacting the economizer rises in the first row, which raises the probability of particle rebound and collision in other rows (compared with Figure 10a–h). For an ash particle diameter of 200 µm, these rebound phenomena occur in all rows. Second place is 150 µm, and 100 µm is the least. The probability of the particle rebound ultimately influences the growth amplification of global erosion. As the ash size increases to a certain numerical value, the probability of the particle rebound reaches the limit, which leads the growth amplification to decrease.

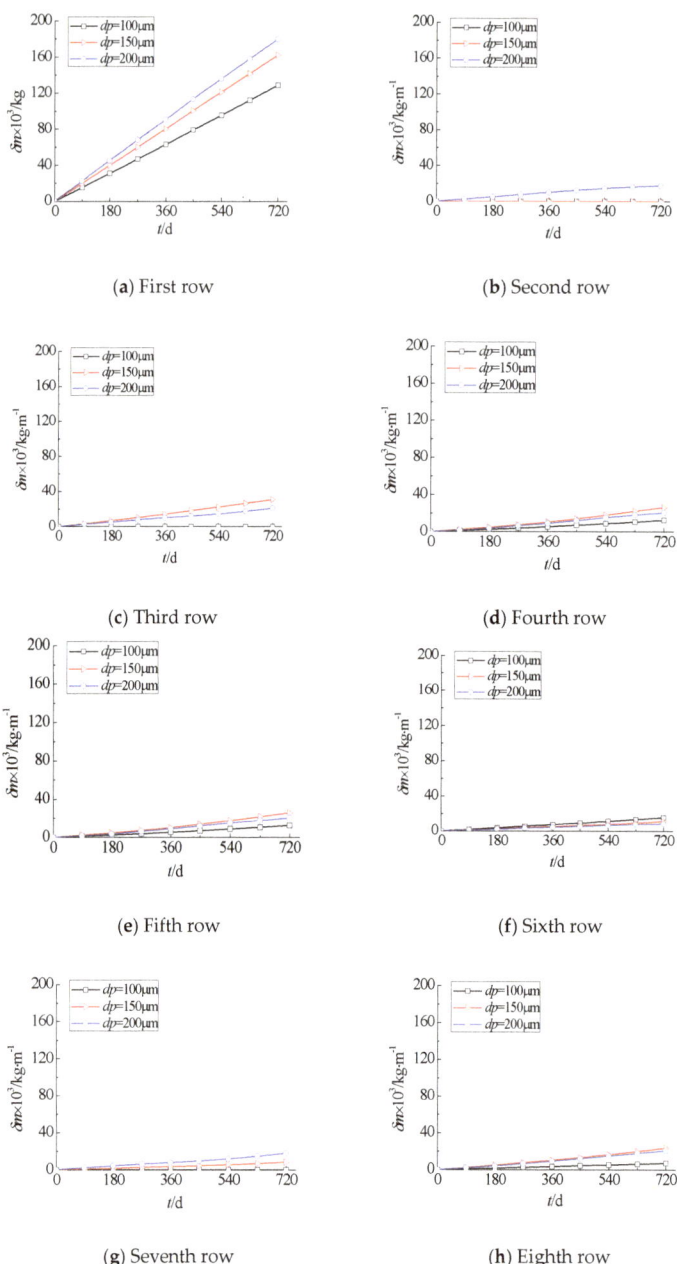

Figure 10. Local erosion loss under different ash sizes.

4.2. Evolution Process of Erosion

4.2.1. Maximum Erosion Depth

Figure 11 shows the maximum erosion depths of the different rows under various ash sizes. The maximum erosion depth is defined as the most critical region, as it represents the thinnest location of the tube wall. From this figure, the maximum erosion depth depends linearly on the working time at the initial time. However, compared to the linear trend (gray dashed line in Figure 11a,b), the growth of the maximum erosion depth slows down gradually. This growing trend forms a favorable flow passage, which prolongs the service life of the economizer effectively.

Since maximum erosion depth in the first row is maximal, this row is the most dangerous and of the greatest concern. Maximum erosion depth of the first row is double or even several times that of other rows. Moreover, not in all rows, the 200 µm ash particle impacts the most powerfully, and the maximum erosion depth is maximal. In the third, fourth, fifth, and eighth row, the maximum is under an ash particle diameter of 150 µm. In the sixth row, the maximum appears under an ash particle diameter of 100 µm. This is mainly caused by particle rebound in the first row, which drives other rows to remove more material as the rebounded particles impact them with different kinetic energies and incidence angles.

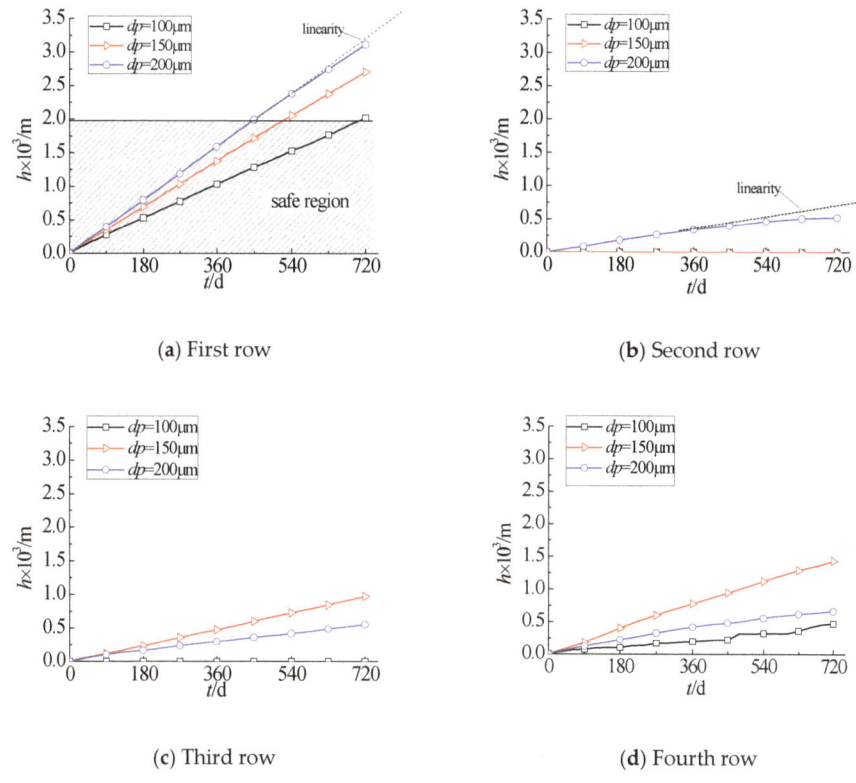

(a) First row

(b) Second row

(c) Third row

(d) Fourth row

Figure 11. *Cont.*

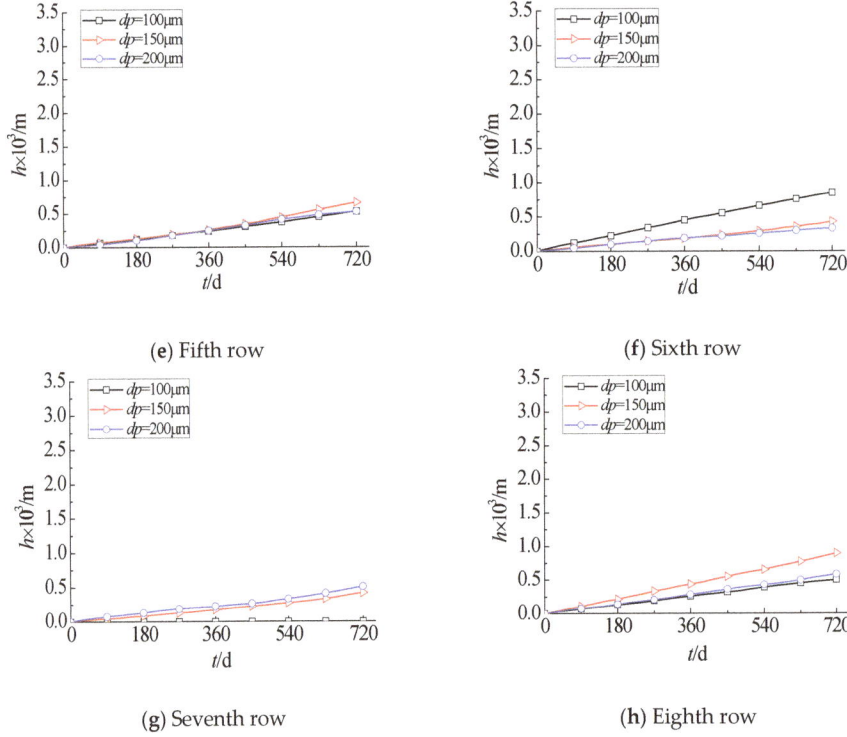

Figure 11. Maximum erosion depth under different ash sizes.

4.2.2. Erosion Profile

Considering that erosion in the first row is the most serious, this paper focuses on the evolution process of the erosion profile for the first row. Figure 12 shows the evolution process of the erosion profile with different ash sizes in the first row. From this figure, as time proceeds, a "V-shape" is gradually formed, which is similar to the erosion profile in Figure 1. This confirms that the dynamic mesh technology is advisable to describe the erosion of economizer tubes.

In addition, θ is introduced as the angle between the position of the erosion profile and the reverse direction of incoming flow, and θ_{max} represents θ in the location of maximum erosion depth in Figure 12a. From Figure 12, larger ash impacts the first row with a smaller angle of θ_{max}. As mentioned earlier, larger ash has greater inertia and less motion, ensuring that the location of the maximum erosion rate appears with a smaller θ.

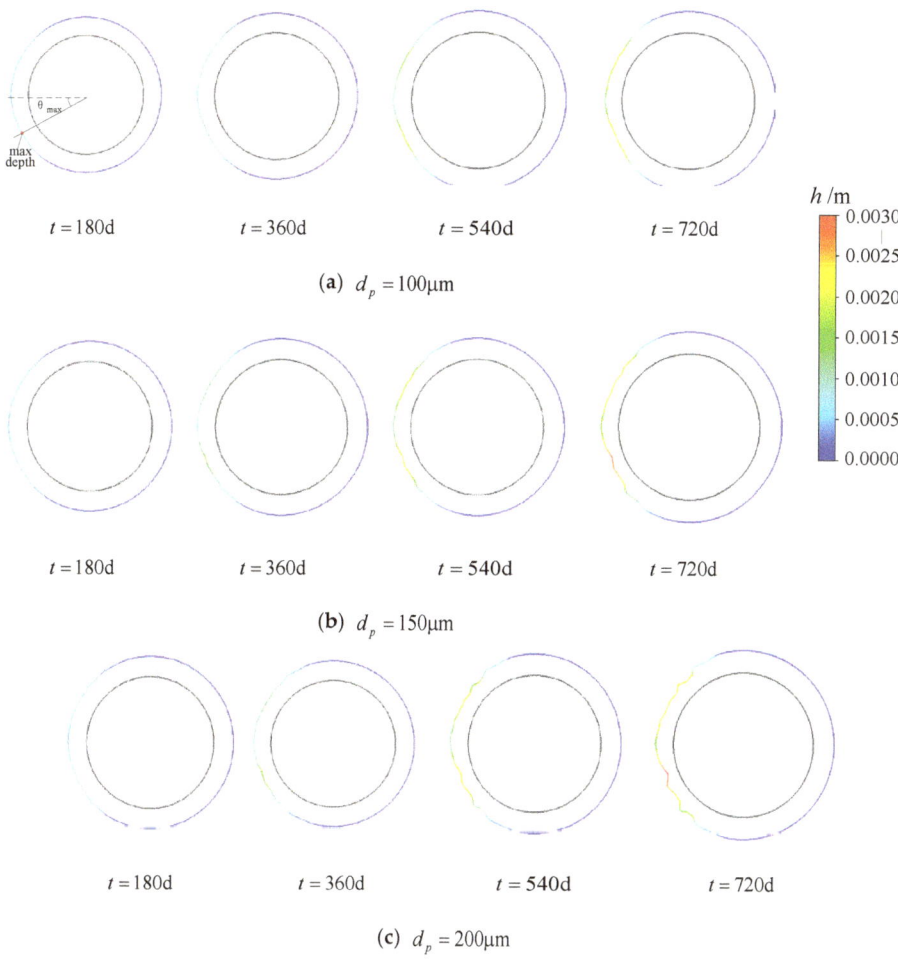

Figure 12. Evolution process of erosion profile in the first row.

4.3. Evolution Process of Particle Motion

Figure 13 shows the particle motion from the specific position in the inlet. From this figure, the particle collision point obviously moves to the center of the economizer tube over time. Meanwhile, due to more material loss with a larger particle, the larger particle moves across a longer distance. In order to describe the particle motion along the radial and tangential direction, Figure 14 is adopted to illustrate the collision point trajectory from the specific position in the inlet, which is the top view of Figure 13a. From this figure, with time increasing, the collision point trajectory moves along the increasing direction of the absolute value of θ. Just because of this, the probabilities of impingement and the growth of the maximum erosion depth slow down gradually in later stages. Although the trend of erosion loss is unlike the maximum erosion depth after 2 years, the erosion loss based on this result is predicted to slowly increase in the future.

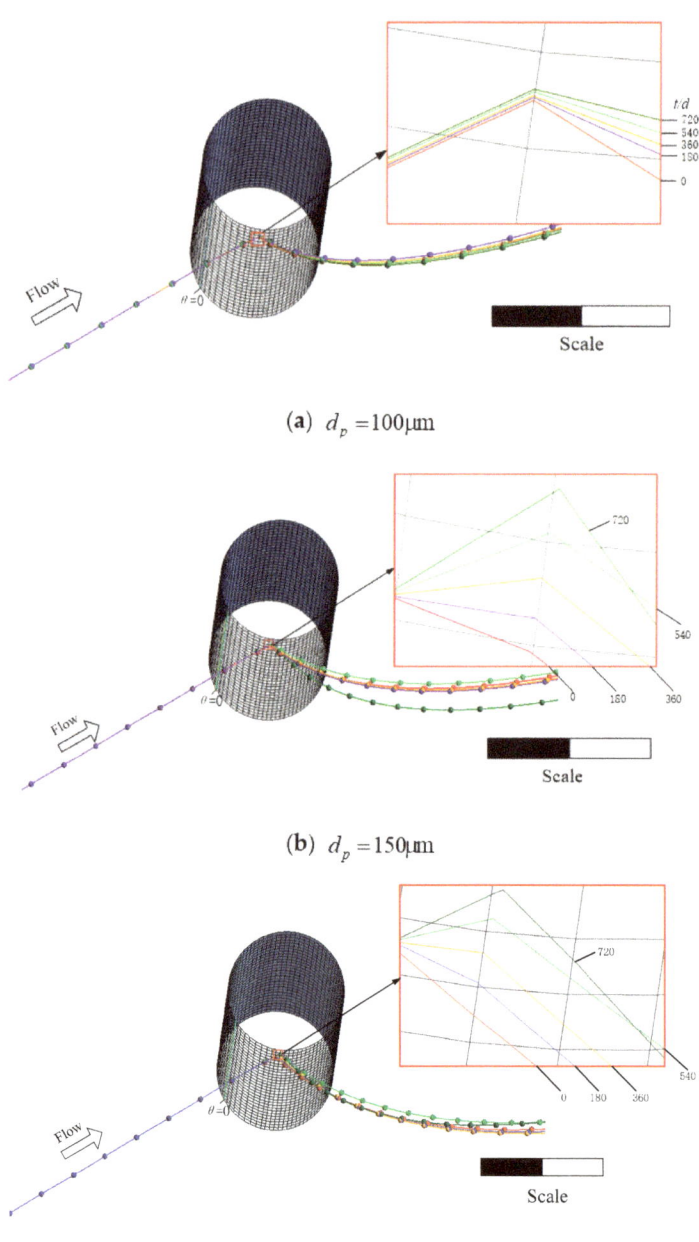

(a) $d_p = 100\mu m$

(b) $d_p = 150\mu m$

(c) $d_p = 200\mu m$

Figure 13. Particle motion.

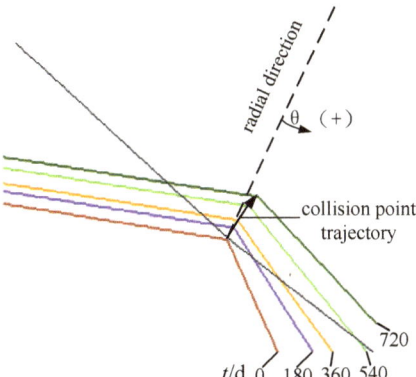

Figure 14. Collision point trajectory ($d_p = 100$ μm).

4.4. Expiry Periods

In Figure 11a, the safe region is defined as the value below 1.99 mm (equal to the critical value of 4.01 mm in wall thickness [25], which is the minimum wall thickness by the specific theoretical calculation). When the maximum erosion depth is up to 1.99 mm, the first row of the economizer tube easily bursts. From this figure, the expiry periods are respectively 710, 530, and 440 days with 100 μm, 150 μm, and 200 μm. These data indicate the expiry period is shortened as the ash diameter increases. Moreover, as the ash size increases, the declining amplification of expiry periods is retarded.

5. Conclusions

The present research aims to investigate a novel method for the solution of the evolution process and its applications to the expiry period of an economizer bank. The CFD-DPM approach coupled with the UDF of the erosion model and the dynamic mesh technology is utilized to solve this complex issue. Based on the results of the applications, the following conclusions are made:

(1) The CFD-DPM approach coupled with the UDF of Huang's erosion model and the dynamic mesh technology can describe the evolution process of erosion on an economizer bank, especially the erosion profile; by comparing the simulation results with the erosion profile on-site, the correctness is verified for this proposed CFD-DPM approach.

(2) The global/local erosion loss and the maximum erosion depth are linearly related to the working time, but the growth of the maximum erosion depth slows down gradually in the later stage; as the ash size increases, the growth amplification of global erosion loss, local erosion in the first row, and the maximum erosion depth decrease.

(3) With increasing time, the collision point trajectory moves along the increasing direction of the absolute value of θ in the first row, which explains why the growth of the maximum erosion depth slows down in later stages.

(4) The expiry period is shortened as the ash diameter increases; moreover, as the ash size increases, the declining amplification of expiry periods is retarded.

Author Contributions: F.S. and X.J. led the project and coordinated the study; Y.D., C.Y. and B.Z. conducted the modelling analysis; Y.D. and Z.Q. conducted the numerical simulation. C.Y. participated in the data analysis. Y.D. and C.Y. prepared, and Z.Q. edited the manuscripts; Y.D. coordinated the manuscript submission. All authors read and approved the final manuscript.

Funding: This research was funded by the National Natural Science Fund Program of China, grant number No. 51706043, and the Jiangsu Natural Science Foundation, grant number No. BK20170679.

Acknowledgments: Research was supported by Datang Nanjing Environmental Protection Technology Co., Ltd. Without the support, this work would not have been possible.

Conflicts of Interest: The authors declare no conflict of interest.

Nomenclature

A_i	area of grid element [m^2]	K	model constant
C	model constant	k_f	fluid thermal conductivity [W/(m·K)]
$C_{1\varepsilon}$	model constant	Kn	Knudsen number
$C_{2\varepsilon}$	model constant	m_p	particle mass [kg]
$C_{3\varepsilon}$	model constant	p	gas pressure [pa]
C_D	drag coefficient	Re	Reynolds number
C_m	model constant	S_k	turbulence kinetic energy source term [kg/(m·s^3)]
C_p	specific heat [J/(kg·K)]	S_ε	turbulence dissipation rate source term [kg/(m·s^4)]
C_S	model constant	T	gas temperature [K]
C_t	model constant	u_f	gas velocity vector [m/s]
C_μ	model constant	u_i	gas velocity of i-direction [m/s]
d_{ij}	deformation tensor	u_p	particle velocity vector [m/s]
d_p	particle diameter [m]	x_i	the coordinate of i-direction [m]
D	model constant	V_p	particle velocity magnitude [m/s]
e	total energy [J/kg]	V_f	gas velocity magnitude [m/s]
e_n	normal coefficient of velocity restitution	Y_M	contribution of the fluctuating dilatation in compressible turbulence to the overall dissipation rate [kg/(m·s^3)]
e_t	tangential coefficient of velocity restitution	α	incidence angle
E_r	ratio of abrasion loss and particle mass [g/kg]	ρ_f	gas density [kg/m^3]
F_D	drag force [N]	ρ_m	density of economizer material [kg/m^3]
F_G	gravity [N]	ρ_p	particle density [kg/m^3]
F_T	thermophoretic force [N]	τ_{ij}	Gas stress tensor [pa]
F_{SL}	*Saffman* lift force [N]	δ_{ij}	*Kronecker* tensor
g	acceleration of gravity [m/s^2]	δ_m	erosion loss [kg/m]
G_b	generation of turbulence kinetic energy due to buoyancy [kg/(m·s^3)]	Δt	time interval [s]
G_k	generation of turbulence kinetic energy due to the velocity gradients [kg/(m·s^3)]	μ	gas viscosity [pa·s]
h	erosion depth [mm]	μ_t	gas second viscosity [pa·s]
$h_{i,j}$	erosion depth of grid element [mm]	σ_k	turbulent *Prandtl* number for k
k_p	particle thermal conductivity [W/(m·K)]	σ_ε	turbulent *Prandtl* number for ε

References

1. Arabnejad, H.; Mansouri, A.; Shirazi, S.A.; McLaury, B.S. Abrasion erosion modeling in particulate flow. *Wear* **2017**, *376–377*, 1194–1199. [CrossRef]
2. Alam, T.; Farhat, Z.N. Slurry erosion surface damage under normal impact for pipeline steels. *Eng. Fail. Anal.* **2018**, *90*, 116–128. [CrossRef]
3. Eichner, D.; Schlieter, A.; Leyens, C.; Shang, L.; Shayestehaminzadeh, S.; Schneider, J.M. Solid particle erosion behavior of nanolaminated Cr$_2$AlC films. *Wear* **2018**, *402–403*, 187–195. [CrossRef]
4. Ji, X.; Wu, J.; Pi, J.; Cheng, J.; Shan, Y.; Zhang, Y. Slurry erosion induced surface nanocrystallization of bulk metallic glass. *Appl. Surf. Sci.* **2018**, *440*, 1204–1210. [CrossRef]
5. William, L. Failure analysis on the economisers of a biomass fuel boiler. *Eng. Fail. Anal.* **2013**, *31*, 101–117. [CrossRef]
6. Finnie, I. Erosion of surfaces. *Wear* **1960**, *3*, 87–103. [CrossRef]
7. Finnie, I. Some reflections on the past and future of erosion. *Wear* **1995**, *186–187*, 1–10. [CrossRef]
8. Bitter, J. A study of erosion phenomena Part I. *Wear* **1963**, *6*, 5–21. [CrossRef]

9. Bitter, J. A study of erosion phenomena Part II. *Wear* **1963**, *6*, 169–190. [CrossRef]
10. Huang, C.; Chiovelli, S.; Minev, P.; Luo, J.; Nandakumar, K. A comprehensive phenomenological model for erosion of materials in jet flow. *Powder Technol.* **2008**, *187*, 273–279. [CrossRef]
11. Grant, G.; Tabakoff, W. *An Experimental Investigation of the Erosion Characteristics of 2024 Aluminum Alloy*; Report; Cincinnati University: Cincinnati, OH, USA, 1973; p. 49.
12. Oka, Y.I.; Okamura, K.; Yoshida, T. Practical estimation of erosion damage caused by solid particle impact: Part 1: Effects of impact parameters on a predictive equation. *Wear* **2005**, *259*, 95–101. [CrossRef]
13. Oka, Y.I.; Yoshida, T. Practical estimation of erosion damage caused by solid particle impact: Part 2: Mechanical properties of materials directly associated with erosion damage. *Wear* **2005**, *259*, 102–109. [CrossRef]
14. Haugen, K.; Kvernvold, O.; Ronold, A.; Sandberg, R. Sand erosion of wear-resistant materials: Erosion in choke valves. *Wear* **1995**, *186–187*, 179–188. [CrossRef]
15. Ahlert, K. Effects of Particle Impingement Angle and Surface Wetting on Solid Particle Erosion of AISI 1018 Steel. Master's. Thesis, The University of Tulsa, Tulsa, OK, USA, 1994.
16. McLaury, B.S. Predicting Solid Particle Erosion Resulting from Turbulent Fluctuations in Oilfield Geometries. Ph.D. Thesis, The University of Tulsa, Tulsa, OK, USA, 1996.
17. Zhang, Y.; Reuterfors, E.P.; McLaury, B.S.; Shirazi, S.A.; Rybicki, E.F. Comparison of computed and measured particle velocities and erosion in water and air flows. *Wear* **2007**, *263*, 330–338. [CrossRef]
18. Shamshirband, S.; Malvandi, A.; Karimipour, A.; Goodarzi, M.; Afrand, M.; Petković, D.; Dahari, M.; Mahmoodian, N. Performance investigation of micro- and nano-sized particle erosion in a 90° elbow using an ANFIS model. *Powder Technol.* **2015**, *284*, 336–343. [CrossRef]
19. Moakhar, R.S.; Mehdipour, M.; Ghorbani, M.; Mohebali, M.; Koohbor, B. Investigations of the failure in boilers economizer tubes used in power plants. *J. Mater. Eng. Perform.* **2013**, *22*, 2691–2697. [CrossRef]
20. Ding, Q.; Tang, X.F.; Yang, Z.G. Failure analysis on abnormal corrosion of economizer tubes in a waste heat boiler. *Eng. Fail. Anal.* **2017**, *73*, 129–138. [CrossRef]
21. Yang, L. One-fluid formulation for fluid–structure interaction with free surface. *Comput. Methods Appl. Mech. Eng.* **2018**, *332*, 102–135. [CrossRef]
22. Joshson, A.A. Dynamic-mesh CFD and its application to flapping-wing micro-air vehicles. In Proceedings of the 25th Army Science Conference, Orlando, FL, USA, 27–30 November 2006; ANSI: Orlando, FL, USA, 2006.
23. Savidis, S.A.; Aubram, D.; Rackwitz, F. Arbitrary *Lagrangian*-eulerian finite element formulation for geotechnical construction processes. *J. Theor. Appl. Mech.* **2008**, *38*, 165–194.
24. Li, J.; Hesse, M.; Ziegler, J.; Woods, A.W. An arbitrary *Lagrangian* Eulerian method for moving-boundary problems and its application to jumping over water. *J. Comput. Phys.* **2005**, *208*, 289–314. [CrossRef]
25. Yuping, C. *Boiler Instruction Manual of DAIHAI in China*; Shang Boiler Works, Ltd.: Shanghai, China, 2006.
26. Versteeg, H.; Malalasekera, W. *An Introduction to Computational Fluid Dynamics: The Finite Volume Method*; Prentice Hall: London, UK, 1995.
27. Gao, W.; Li, Y.; Kong, L. Numerical investigation of erosion of tube sheet and tubes of a shell and tube heat exchanger. *Comput. Chem. Eng.* **2017**, *96*, 115–127. [CrossRef]
28. Launder, B.E.; Spalding, D.B. The numerical computation of turbulent flows. *Comput. Methods Appl. Mech. Eng.* **1974**, *3*, 269–289. [CrossRef]
29. ANSYS. *Ansys Fluent User's Guide*; Ansys Inc.: Canonsburg, PA, USA, 2015.
30. Grant, G.; Tabakoff, W. Erosion Prediction in Turbomachinery Resulting from environmental solid particles. *J. Aircr.* **1975**, *12*, 471–478. [CrossRef]
31. ASTM. Standard test method for conducting erosion tests by solid particle impingement. *ASTM Int.* **2013**. [CrossRef]
32. Sagayaraj, T.A.D.; Suresh, S.; Chandrasekar, M. Experimental studies on the erosion rate of different heat treated carbon steel economizer tubes of power boilers by fly ash particles. *Int. J. Miner. Metall. Mater.* **2009**, *16*, 534–539. [CrossRef]

© 2018 by the authors. Licensee MDPI, Basel, Switzerland. This article is an open access article distributed under the terms and conditions of the Creative Commons Attribution (CC BY) license (http://creativecommons.org/licenses/by/4.0/).

Article

From a Sequential Chemo-Enzymatic Approach to a Continuous Process for HMF Production from Glucose

Alexandra Gimbernat [1], Marie Guehl [2,3], Nicolas Lopes Ferreira [4], Egon Heuson [1], Pascal Dhulster [1], Mickael Capron [2], Franck Dumeignil [2], Damien Delcroix [3], Jean-Sébastien Girardon [2] and Rénato Froidevaux [1,*]

[1] Univ. Lille, INRA, ISA, Univ. Artois, Univ. Littoral Côte d'Opale, EA7394–ICV-Institut Charles Viollette, F-59000 Lille, France; alexandra.gimbernat@gmail.com (A.G.); egon.heuson@univ-lille.fr (E.H.); pascal.dhulster@univ-lille.fr (P.D.)

[2] Univ. Lille, CNRS, Centrale Lille, ENSCL, Univ. Artois, UMR 8181-UCCS-Unité de Catalyse et Chimie du Solide, F-59000 Lille, France; guehl.marie@gmail.com (M.G.); mickael.capron@univ-lille.fr (M.C.); franck.dumeignil@univ-lille.fr (F.D.); jean-sebastien.girardon@univ-lille.fr (J.S.G.)

[3] IFP Energies Nouvelles, Rond-point de l'échangeur de Solaize, BP 3, 69360 Solaize, France; damien.delcroix@ifpen.fr

[4] IFP Energies Nouvelles, 1 et 4 avenue de Bois-Préau, 92852 Rueil Malmaison, France; nicolas.lopes-ferreira@ifpen.fr

* Correspondence: Renato.froidevaux@univ-lille.fr; Tel.: +33-320-417-566

Received: 13 July 2018; Accepted: 27 July 2018; Published: 17 August 2018

Abstract: Notably available from the cellulose contained in lignocellulosic biomass, glucose is a highly attractive substrate for eco-efficient processes towards high-value chemicals. A recent strategy for biomass valorization consists on combining biocatalysis and chemocatalysis to realise the so-called chemo-enzymatic or hybrid catalysis. Optimisation of the glucose conversion to 5-hydroxymethylfurfural (HMF) is the object of many research efforts. HMF can be produced by chemo-catalyzed fructose dehydration, while fructose can be selectively obtained from enzymatic glucose isomerization. Despite recent advances in HMF production, a fully integrated efficient process remains to be demonstrated. Our innovative approach consists on a continuous process involving enzymatic glucose isomerization, selective arylboronic-acid mediated fructose complexation/transportation, and chemical fructose dehydration to HMF. We designed a novel reactor based on two aqueous phases dynamically connected via an organic liquid membrane, which enabled substantial enhancement of glucose conversion (70%) while avoiding intermediate separation steps. Furthermore, in the as-combined steps, the use of an immobilized glucose isomerase and an acidic resin facilitates catalyst recycling.

Keywords: (bio) catalysis; biomass; glucose; 5-hydroxymethylfurfural (HMF); chemo-enzymatic catalysis

1. Introduction

In a context of the fast depletion of fossil resources, lignocellulosic biomass possesses a high potential as a sustainable raw material for fuels [1] and fine chemicals [2,3] production in industrially significant volumes. Upon specific physical, chemical and/or biological treatments of lignocellulose, access can be gained to monomeric sugars [4] like glucose as its main saccharidic component. Glucose is, as a result, currently considered as one of the most promising starting materials for the conversion of biomass into drop-in or new high-value platform chemicals [5].

Efficient and eco-responsible valorization routes for this new renewable carbon source are highly desirable, and every field of catalysis should be mobilized as the key to success for a selective valorization of such a polyfunctional substrate. Biocatalytic processes are already or are about to be industrialized [6] and heterogeneous or homogeneous catalysis, alone or in combination, have shown high efficiency for biomass valorization, but lack of selectivity still causes problems in the products' separation steps [7]. In the last few years, a new paradigm of catalysis has emerged to answer the challenges of selectivity and productivity by combining biocatalysis and chemocatalysis [8–10]. This combination is known as chemo-enzymatic catalysis or hybrid catalysis [11,12].

In the field of sustainable and biosourced chemistry, glucose conversion to 5-hydroxymethylfurfural (HMF) deserves specific attention as a platform molecule [13]. Its hydrogenation, for instance, can lead to 2,5-dimethylfuran (2,5-DMF), a new fuel additive [14] and a precursor of terephthalic acid after dehydrative Diels–Alder reaction with ethylene to p-xylene and subsequent oxidation [15,16]. HMF oxidation is also envisioned to obtain 2,5-furandicarboxylic acid (FDCA), a biosourced alternative to terephthalic acid in polyester plastic production with a high market potential [17,18]. Unfortunately, abundant glucose is not the substrate of choice for efficient dehydration to HMF compared to its more expensive fructose isomer. The latter presents a fructofuranose mutamer more prone to dehydration to a furanic ring with lower energetic barriers [19]. Isomerization of glucose to fructose thus appears as a key step that can hardly be circumvented. A well-known transformation of glucose to fructose is the enzymatic glucose isomerization, employed in the industrial production of high-fructose corn syrup "HFCS" [20]. Despite the need for expensive high-purity glucose, for buffers and multiple ion-exchange resins to get rid of all the metallic residues in alimentary HFCS, enzymatic isomerization remains the preferred process compared to chemical isomerization, even if the latter receives a recent renewed interest [21–25]. The aforementioned enzymatic reaction reaches a thermodynamic equilibrium (Keq~1), which limits glucose conversion [26–28]. Whereas this equilibrium is particularly well-adapted for HFCS production involving glucose/fructose mixtures, HMF synthesis requires a pure fructose feed for the subsequent dehydration step, where compatibility issues between catalysts arise: Huang et al. indeed shown incompatibility between a thermophilic glucose isomerase immobilized on aminopropyl-functionalized mesoporous silica (FMS) and a heterogeneous Brønsted acid propylsulfonic-FMS-SO$_3$H in a THF:H$_2$O (4:1 v/v) mixture [29]. Isomerization is carried out at 363 K to reach 61% fructose yield, and then the temperature has to be increased to 403 K to reach a 30% HMF yield. In the meantime, glucose isomerase is fully denatured.

Strategies have thus been adopted to combine enzymatic isomerization of glucose to fructose and chemical by separating bio- and chemo-catalysis involving fructose transportation between the isomerization aqueous phase and an organic phase for subsequent reactivity. Huang et al. [30] describe this concept for the first time by adding sodium tetraborate in the aqueous isomerization medium to form a fructoborate compound by complexation with fructose. Transportation of this complex to a separated organic phase is assisted by a cationic quaternary ammonium and enables enhanced glucose conversion by shifting the isomerization equilibrium towards fructose. The separate dehydration of fructose in the organic phase gives an increase in HMF yield up to 63% compared to a yield of 28% HMF without borate addition, associated to a glucose conversion of 88% instead of 53%. However, in this work, the complexation selectivity between glucose and fructose was not satisfactory and was thus further improved by Palkovits et al. who optimized the chemical nature of the complexing boronate species [31]. A global process exploiting this concept for glucose to HMF production has concomitantly been proposed by Alipour et al. [32] Therein, fructose is complexed with phenylboronic acid and transferred to an organic phase, which is separated and contacted with an acidic ionic liquid phase to promote the release of free fructose. This fructose-rich ionic liquid is further used for dehydration to HMF in a biphasic medium. The produced HMF is then back-extracted to a last low-boiling point organic phase. This method involves the use of four different media with intermediate phase separations. Inspired by these sequential processes, we propose here to circumvent this drawback by setting up an innovative integrated implementation minimizing the number of steps and preventing, among other things, the use of costly and hard-to-recover ionic liquids.

Herein, we highlight an efficient and fully-integrated cascade reactions process to convert glucose to HMF using an organic liquid membrane separating two aqueous phases for fructose transportation. This approach that we first proposed as a concept exemplifying the potential of hybrid catalysis [33] is based on continuous aqueous glucose isomerization, fructose complexation, extraction and transportation towards an intermediate organic phase. Fructose is then released at the interface with the second aqueous phase (aqueous receiving phase), where it is subsequently dehydrated to HMF. This integrated process without intermediate phase recovery, separation and recycling is presented in Figure 1. Before fully designing the entire process, where all the reactions are carried out simultaneously, we separately investigated each reaction to notably determine their respective optimal ranges of conditions before combination: (1) glucose isomerization to fructose occurs in a primary aqueous feed phase; (2) the as-formed fructose transportation is then made possible through selective complexation with phenylboronic acid as a carrier associated to a quaternary ammonium into an organic solvent (4-methyl-2-pentanone "MIBK"); (3) eventually, the fructoboronate complex is hydrolyzed by acidic conditions in a receiving aqueous phase containing an acidic resin, which also promotes dehydration of fructose to HMF. MIBK was chosen among different solvents, such as methyl-*tert*-amyl-ether, 5-methyl-2-hexanone, dimethyl carbonate, etc. ... but the study is not presented here. Our criteria were the solubility, toxicity, boiling point. With MIBK and methyl-*tert*-amyl-ether, we have obtained the best results in terms of fructose extraction yield but we have chosen MIBK (117 °C) due to its higher boiling point than that of methyl-*tert*-amyl-ether (86 °C). This criterion was important for the implementation of the chemical catalysis step at 80–90 °C. Finally, our approach was not only focused on the compatibility issues of bio- and chemocatalytic reactions with the aim of overcoming the isomerization equilibrium limitation, but also pays particular attention to integrating all the steps to minimize separation and recycling burdens, which can be detrimental for the overall economics and efficiency of the process. The methodology envisioned to move from a sequential approach towards an integrated continuous process is schematized in Figure 2.

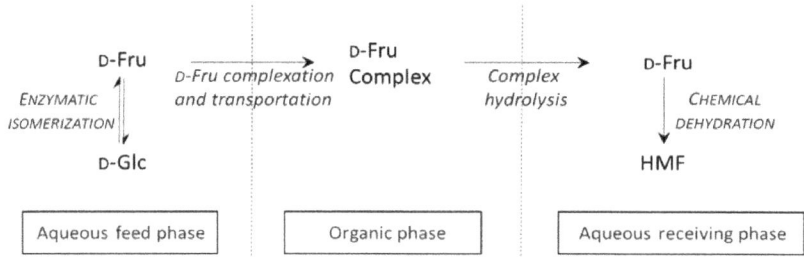

Figure 1. Hybrid catalysis simultaneous process applied to the transformation of glucose to HMF (D-Glc = D-glucose, D-Fru = D-Fructose, HMF = 5-hydroxymethylfurfural).

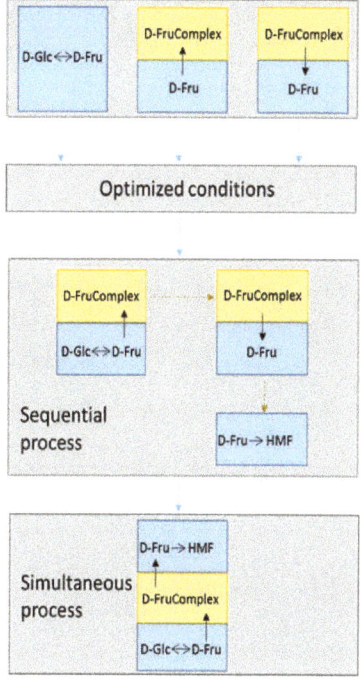

Figure 2. Methodology from a sequential approach towards an integrated continuous process.

2. Results and Discussion

2.1. Study of D-Glucose Enzymatic Isomerization in the Aqueous Feed Phase

The optimal temperature of use ranges from 333 to 353 K and the optimum pH ranges from pH 7.0 to 9.0 [34]. We first checked the influence of pH and temperature using the lot of Immobilized Glucose Isomerase (IGI) received from our supplier. A glucose solution at pH 7.5 (Tris-HCl buffer, 100 mM) was prepared and isomerized from 323 K to 363 K. The relative enzymatic activity is presented in Figure 3, which highlights a maximal enzymatic activity at 343 K. A glucose solution of a pH from 4.5 to 9 (Tris-HCl buffer, 100 mM from pH 7 to 9; sodium phosphate buffer, 100 mM pH 6.5; sodium citrate buffer, 100 mM pH 4.5) was thus prepared and isomerized at 343 K. The relative enzymatic activity presented in Figure 4 shows a maximal IGI activity at a pH around 7.5. It is noteworthy that the IGI maintained more than 80% of its optimal activity between pH 7 and 9. Below a pH of 6.5, the IGI lost more than 50% of its activity. The pH and the temperature chosen must both allow the optimal functioning of the enzyme but also promote the extraction phenomenon by maintaining a pH larger than the pKa of boronic acid. Indeed, the extraction of D-fructose by formation of a complex [D-fructose-boronic acid]$^-$ is favored if the pH of the aqueous solution containing D-fructose is larger than the pKa of boronic acid [22]. The pKa of the main boronic acids are less than 8.5. In this context, we chose pH 8 that makes it possible to overcome the pKa-related lock of the selected boronic acid to transport fructose to the organic phase while maintaining a relative activity of more than 80% of IGI. A D-glucose enzymatic isomerization was then performed in the selected conditions (Tris-HCl 100 mM, pH 8.5, 343 K). The results are presented in Figure 5 as a function of time. A blank reaction (grey dots) was run in the absence of IGI and showed the absence of D-glucose isomerization or degradation in the chosen experimental conditions. When IGI was present in the system, the D-fructose amount increased from 0% to 42% during the first 30 min. From 30 to 60 min, the D-glucose conversion was slower, while the D-Fructose amount increased from 42% to 55%. A maximum D-glucose conversion of 55% was finally reached in agreement with the glucose-fructose thermodynamic equilibrium expected at this temperature (Keq = 1.23), as already observed by McKay et al. [35].

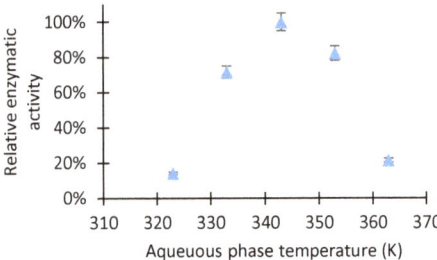

Figure 3. Relative enzymatic activity of IGI in D-glucose isomerization at pH 7.5 as a function of temperature.

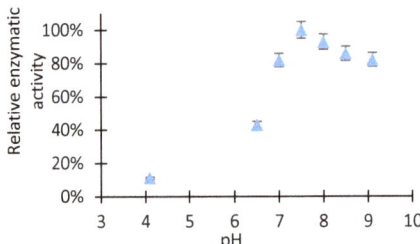

Figure 4. Relative enzymatic activity of IGI in D-glucose isomerization at 343 K as a function of pH.

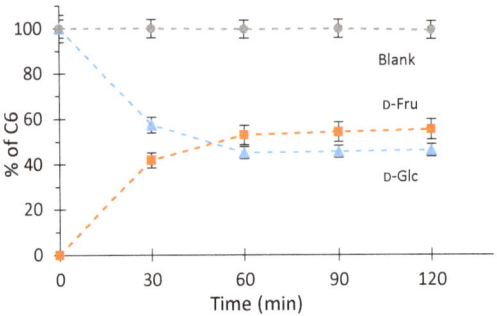

Figure 5. D-glucose consumption during the isomerization step as a function of time; (Δ), D-fructose formation (□), D-glucose amount in the blank experiment at pH 8.5 (○). Initial conditions: D-Glucose (100 mM), IGI (0.5 g), Tris-HCl 100 mM, pH 8.5, V = 100 mL, 343 K, 750 rpm.

2.2. Organic Liquid Membrane for D-Fructose Complexation/Transportation

In order to enhance the isomerization yield, simultaneous isomerization and complexation/transportation of D-fructose in a liquid/liquid aqueous/organic biphasic system was further studied. The temperature was set at 343 K and the pH was chosen as a trade-off to simultaneously maximize the enzymatic activity (7 < pH < 9); to avoid monosaccharide degradation, and to enhance the extraction process as reported previously [36], the pH was thus kept at a value of

8.5 by using a Tris-HCl 100 mM buffer. The extraction was conducted in an organic solvent with low water miscibility (MIBK) containing a lipophilic arylboronic acid (carrier) and a quaternary ammonium salt (Aliquat336®) as a phase transfer agent [37]. At pH 8.5, aryl boronic acid $ArB(OH)_2$ was actually present under its hydroxylated anionic form, as a tetrahedral aryltrihydroxyborate $ArB(OH)_3^-$ [38]. At the interface between the aqueous and organic phases, D-fructose further reacts with the arylborate to form a tetrahedral fructoboronate ester (Figure 6) [38]. The fructoboronate complex then forms an intimate ion pair with Aliquat336®, which enables its transportation to the organic phase [39].

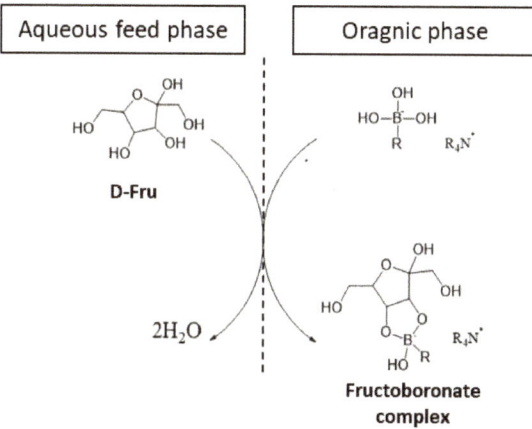

Figure 6. Fructoboronate complex formation at the interphase between the first aqueous phase and the organic liquid membrane.

2.2.1. Influence of the Boronic Acid Structure

The influence of the boronic acid structure was investigated in order to optimize kinetics and maximize the selectivity of D-fructose complexation/transportation. Seven arylboronic acids differing by the electronic properties of their substituents and thus by their pKa were screened in the complexation reaction with D-fructose (Figure 7): 2,3-DCPBA, 2,4-DCPBA, 3,4-DCPBA, 3,5-DCPBA, PBA, 4-TBPBA, 4-TFMPBA. The complexation/transportation, which will be referred to as "extraction" in the following sections, was carried out in a biphasic system, as described before. The pKa of the different boronic acids and the relative extraction yields are summarized in Table 1 and reported in Figure 8.

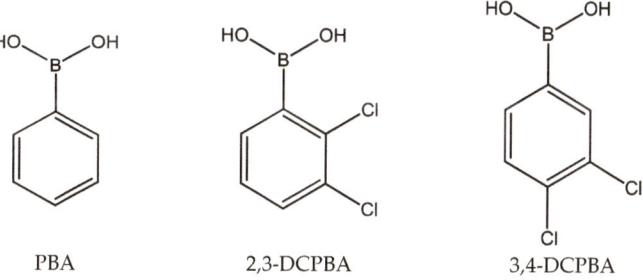

Figure 7. *Cont.*

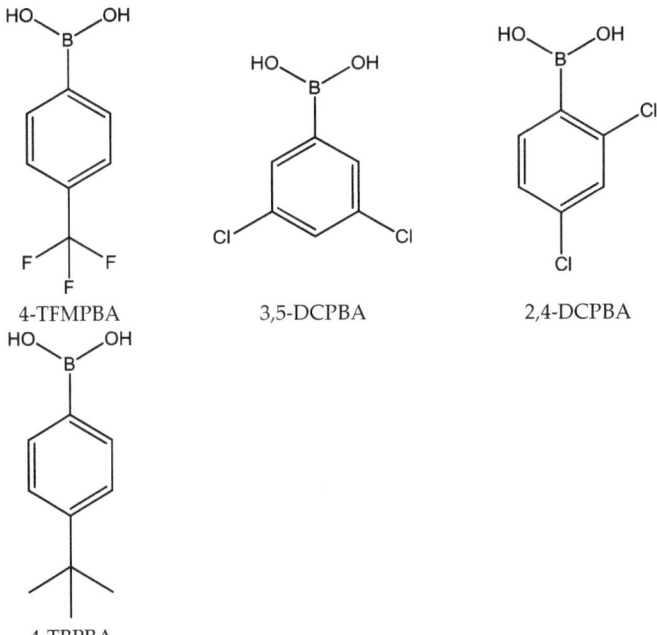

Figure 7. Structures of the arylboronic acids used for extraction of fructose.

Table 1. Influence of the boronic acid structure on the extraction yield and rate.

Boronic Acid	pKa	Extraction Yield %		Extraction Rate µmol/min	
4-TBPBA	9.3	8.3	± 5.6	0.52	± 0.08
PBA	9.1	32.4	± 0.3	1.27	± 0.39
2,4-DCPBA	8.9	43.3	± 1.6	1.56	± 0.34
3,4-DCPBA	7.4	46.5	± 4.9	1.48	± 0.23
2,3-DCPBA	7.4	49,2	± 1.6	1.99	± 0.17
4-TFMPBA	9.1	50.3	± 2.2	1.26	± 0.12
3,5-DCPBA	7.4	55.3	± 0.9	1.94	± 0.20

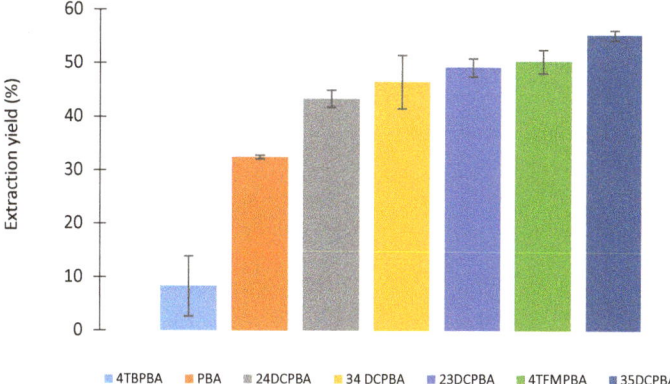

Figure 8. Influence of the arylboronic acid structure on the complexation/transportation process after 45 min. [D-Fructose]$_i$ = 100 mM, [Boronic acid] = 100 mM, [Aliquat336®] = 200 mM, Tris HCl 100 mM pH 8.5, MIBK, 343 K, 750 rpm.

A group composed by 2,4-DCPBA, 3,4-DCPBA, 2,3-DCPBA, 4TFMPBA and 3,5-DCPBA showed the best results with extraction yields between 43% and 55%, while with 4-TBPBA and PBA only 8% and 32% were achieved, respectively. These results also show that the D-fructose extraction yield increases when the pKa of the boronic acid used decreases. Better extraction yields were observed when the pKa of the boronic acid was lower than (or close to) the pH of the aqueous feed phase (8.5). This analysis supports the hypothesis from Morin et al. [36] and Karpa et al. [37] indicating that higher extraction yields are obtained when pH ≥ pKa, favoring a predominant tetrahedral anionic borate form of the complex, which enables its transportation by association with the ammonium carrier. An exception was, however, observed with 4-TFMPBA with a pKa of 9.1. This might be due to the fact that 4-TFMPBA has a higher dipolar moment (2.90 D, against 0.43 D to 1.86 D for the other used boronic acids) that could increase the reactivity. The extraction rates and the pKas (Table 1) seem to be correlated. A higher pKa of the boronic acid was linked with a lower initial extraction rate. Among the most effective boronic acids, 3,4-DCPBA was selected for further studies.

2.2.2. Influence of the Boronic Acid: Aliquat336® Molar Ratio

With 3,4-DCPBA as the selected boronic acid, the influence of the molar ratio between the boronic acid and the ammonium salt Aliquat336® in the organic solvent was investigated. Figure 9 shows the evolution of D-Fructose concentration in the aqueous phase as a function of time for different Aliquat336® concentrations (from 0 to 400 mM). D-Fructose and 3,4-DCPBA were initially introduced at concentrations of 100 mM. In the first 15 min of reaction, the concentration of D-Fructose decreased irrespective of the Aliquat336® concentration with various initial rates. An equilibrium seemed to be progressively reached between 15 and 45 min, as no more evolution of the D-Fructose concentration was then observed. A blank reaction (blue dots) without Aliquat336® nor 3,4-DCPBA was also tested, which showed no evolution. From these observations, it can be considered that the molar ratio 3,4-DCPBA:Aliquat336® is a relevant parameter to control the initial D-fructose extraction rate. Figure 10 shows the evolution of the initial extraction rate and of the D-fructose extraction yield for different 3,4-DCPBA:Aliquat336® molar ratios (from 1:0.5 to 1:4). The initial extraction rate continuously increased when the 3,4-DCPBA:Aliquat336® molar ratio increased from 1:0.5 to 1:2 with respectively 0.37 µmol/min and 1.50 µmol/min. Above a 1:2 molar ratio, no further kinetic benefits could be clearly observed from such an excess of Aliquat336® in the reaction medium. This result is consistent with an ion-pairing phenomenon between the fructoboronate complex and the Aliquat336® occurring at the interphase enabling D-fructose transportation. The D-fructose extraction yield increased when the

3,4-DCPBA:Aliquat336® molar ratio increased from 1:0.5 to 1:2 with respectively 21.5% and 50.0%. As in the case of the initial extraction rate, above a 1:2 molar ratio no further benefits on the extraction yield were displayed by using an excess of Aliquat336®. Therefore, in order to maximize the initial extraction yield and the initial rate while optimizing the quantity of ammonium salt added, an arylboronic acid:Aliquat336® ratio of 1:2 was selected for the following studies.

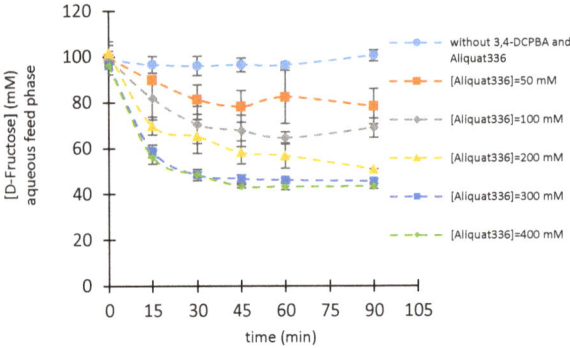

Figure 9. Evolution of [D-fructose] in the aqueous phase as a function of time for different 3,4-DCPBA/Aliquat336® ratios. [D-fructose]$_i$ = 100 mM, [3,4-DCPBA] = 100 mM, [Aliquat336®] = X mM (X varies from 0 to 400 mM), Tris-HCl 100 mM pH 8.5, MIBK, 343 K, 750 rpm.

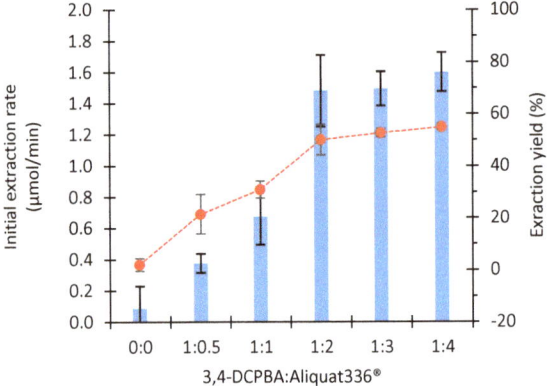

Figure 10. Initial extraction rate in aqueous phase (blue rods) and extraction yield (red dots) in function of 3,4-DCPBA:Aliquat336® molar ratio. [D-Fructose]$_i$ = 100 mM, [3,4-DCPBA] = 100 mM, [Aliquat336®] = X mM (X varies from 0 to 400 mM), Tris-HCl 100 mM, pH 8.5, MIBK, 343 K, 750 rpm.

2.2.3. Influence of Boronic Acid and Aliquat336® Concentrations

Keeping a molar ratio of arylboronic acid: Aliquat336® equal to 1:2 and at fixed initial D-Fructose amount, both 3,4-DCPBA and Aliquat336® concentrations were varied, considering the best compromise between component concentrations and costs (Figure 11). From 0 to 15 min, the D-Fructose extraction yield increases rapidly and then stabilizes for lowest 3,4-DCPBA concentrations less than 100 mM. The best extraction yields are thus obtained for higher 3,4-DCPBA concentrations, the final yield being the highest at 300 mM. The slight difference in extraction yields between 100 mM and 300 mM 3,4-DCPBA concentration does not justify the use of such a concentrated solution. From Figure 12, the initial extraction rate values for different 3,4-DCPBA:Aliquat336® concentrations have been calculated (Table 2). Two behaviors can be observed, the first one for the 25:50 ratio and the second one for all the other ratios, namely from 50:100 to 300:600. We observed that the larger the 3,4-DCPBA amount, the higher the initial extraction rate. However, after a certain amount of 3,4-DCPBA, the interface seems to be saturated by 3,4-DCPBA:Aliquat336® complexes and then the initial rate does not increase anymore. Therefore, further optimization steps considered a concentration of 100 mM while keeping a molar ratio of 1:2 with Aliquat 336®.

Table 2. Initial extraction rates for various concentrations in the system. [D-Fructose]$_i$ = 100 mM, [3,4-DCPBA] = Y mM (Y varies from 25 to 300 mM), [Aliquat336®] = Yx2 mM, Tris-HCl 100 mM, pH 8.5, MIBK, 343 K, 750 rpm.

[3,4-DCPBA] (mM)	Extraction Rate (μmol/min)	
25	0.89	±0.15
50	1.37	±0.41
150	1.45	±0.05
100	1.22	±0.15
200	1.40	±0.21
300	1.33	±0.14

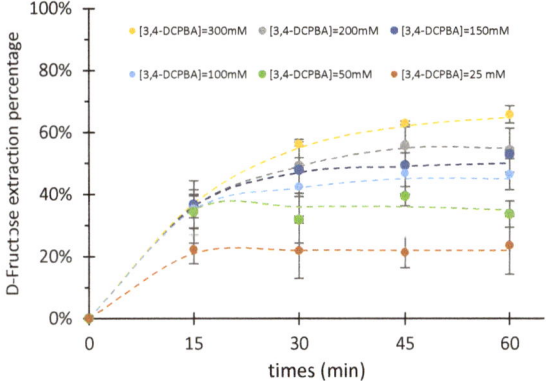

Figure 11. Evolution of the D-Fructose extraction yield for different 3,4-DCPBA and Aliquat336® concentrations for a 1:2 molar ratio. [D-Fructose]$_i$ = 100 mM, [3,4-DCPBA] = Y mM (Y varies from 25 to 300 mM), [Aliquat336®] = Y × 2 mM, Tris-HCl 100 mM, pH 8.5, MIBK, 343 K, 750 rpm.

2.2.4. Influence of the D-Fructose: Boronic Acid Molar Ratio

The influence of the D-Fructose: 3,4-DCPBA molar ratio was further investigated. Figure 12 shows the D-fructose extraction yield as a function of this molar ratio. The yield ranged between 40% and 60%, for ratios from 0.25:1 to 1:1 and between 18% and 28% for ratios from 2:1 to 10:1. When the 3,4-DCPBA was introduced in excess or stoechiometrically compared to D-Fructose (red rods), the extraction yield was larger than 40%, reaching a maximum of 60% for a two-fold excess of boronic acid. However, when it was introduced in default (blue rods), the extraction yield did not exceed 35% without any significant further evolution between a molar ratio of 2:1 to 10:1.

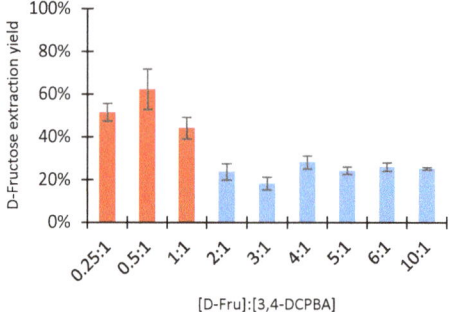

Figure 12. Influence of the initial D-Fructose:3,4-DCPBA molar ratio on the the extraction yield. [D-Fructose]$_i$ = X mM, [3,4-DCPBA] = 100 mM, [Aliquat336®] = 200 mM, Tris-HCl 100 mM, pH 8.5, MIBK, 343 K, 750 rpm.

2.2.5. Influence of D-fructose and Boronic Acid Concentrations

The evolution of the extracted D-fructose concentration as a function of the initial D-fructose concentration, for the same 3,4-DCPBA concentration, is shown in Figure 13. Logically, when the initial concentration of D-fructose increased from 25 mM to 1000 mM, the extracted concentration linearly increased, according to a straight correlation between the initial amount of D-fructose and the amount of extracted D-fructose. Keeping a molar D-fructose:boronic acid ratio equal to 1:1 and a molar ratio of arylboronic acid:Aliquat336® equal to 1:2, a benchmark experiment was set up with the following parameters: T = 34 K, [D-Fructose]$_i$ = 100 mM, [3,4-DCPBA] = 100 mM, [Aliquat 336®] = 200 mM. The other experiments have been set up by doubling all concentrations. Table 3 presents the extraction yields and initial extraction rates depending on the variation of the initial D-fructose concentration while keeping all the molar ratios relative to 3,4-DCPBA and Aliquat336® as constant. An increase in the concentrations (D-Fructose, 3,4-DCPBA and Aliquat336®) led to a decrease in the D-Fructose extraction yield. Indeed, at 100 mM of D-Fructose, 43.3% of the fructoboronate complex were extracted whereas only 32.7% and 27.3% were respectively extracted at 200 mM and 300 mM of D-fructose. Therefore, the optimal conditions ([D-Fru]/[3,4-DCPBA]/[Aliquat336] ratio of 100/100/200 mM) were kept for the following studies.

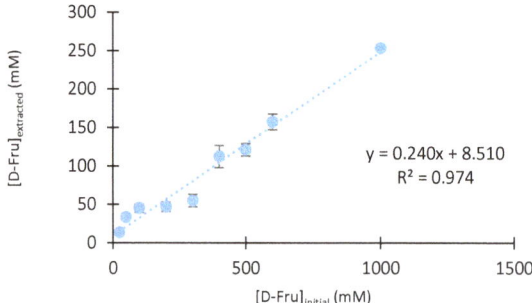

Figure 13. Influence of the initial D-Fructose concentration on the amount of D-fructose extracted. [D-Fructose]$_i$ = X mM (X varies from 25 to 1000), [3,4-DCPBA] = 100 mM, [Aliquat336®] = 200 mM, Tris-HCl 100 mM, pH 8.5, MIBK, 343 K, 750 rpm.

Table 3. D-fructose extraction yield and initial extraction rate for different concentrations.

[D-Fru] mM	[3,4-DCPBA] mM	[Aliquat336®] mM	Extraction Yield %	
100	100	200	43.3	± 4.22
200	200	400	32.6	± 1.22
300	300	600	27.3	± 4.6

2.2.6. Hydrolysis of the Fructoboronate Complex for D-Fructose Release in the Second Aqueous Phase

The further stage of the D-fructose transportation is the fructoboronate complex hydrolysis for D-fructose release in the second aqueous phase (the aqueous receiving phase). This hydrolysis step allows the release of D-fructose in this phase. The influent parameter for the hydrolysis reaction is the pH of the aqueous solution which will be the "reservoir" of the H^+ ions necessary for hydrolysis of the fructoboronate complex. Therefore, the influence of the pH of the second aqueous phase on the release yield was investigated. pHs of 1, 3, 5 and 8 were tested by using with H_2SO_4 for pH 1, citrate buffer for pH 3 and 5 and Tris HCl buffer for pH 8. Prior to these experiments, a D-fructose extraction experiment was performed complexation/transportation in the optimized conditions (100 mM D-fructose and a D-fructose: 3,4-DCPBA:Aliquat336® molar ratio of 1:1:2). The organic phase was then recovered and contacted with the second aqueous phase for the release step. As observed by Paugam et al. [40] and keeping in mind the final objective of this work that is the obtaining of a simultaneous process, the saccharide flow from the basic aqueous donor phase to the acidic receiving phase through an organic phase is favored when a pH gradient is applied between the two phases. Figure 14 shows the evolution of the D-fructose release yield as a function of time for different pH of the aqueous receiving phase. In the cases of pH 8 and pH 1, an increase was observed until 30 min (with an initial rate of about 0.6 and 1.8 μmol/min, respectively, Table 4) and a steady-state was reached for a yield of 22.7% and 54.5% for pH 8 and pH 1, respectively. Concerning pH 3 and pH 5, an increase was observed up to 60 min (with an initial rate of 1.5 μmol/min, Figure 14) to obtain respectively 91% and 100% of release. Considering that the pKa of 3,4-DCPBA is around 7.4, when the pH of the receiving aqueous phase is 8, the pH is then superior to pKa. Then, the release mechanism is unfavorable [40]. With pH 3 and 5, more than 90% of the extracted D-Fructose were released. Moreover, the experiment with H_2SO_4 at pH 1 shows full hydrolysis of the fructoboronate complex followed by conversion of D-Fructose to HMF, levulinic and formic acid, which explains the high release rate and the decrease in the actual D-fructose concentration after 15 min.

Table 4. Transport yield and extraction rate of D-fructoboronate hydrolysis to D-fructose and D-fructose release in the aqueous phase as a function of its pH.

pH	Transport Yield (%)		Extraction Rate (μmol/min)	
5	100	± 4.2	1.58	± 0.14
3	91.5	± 6.3	1.53	± 0.13
1	54.5	± 3.7	1.84	± 0.09
8	22.7	± 5.2	0.63	± 0.14

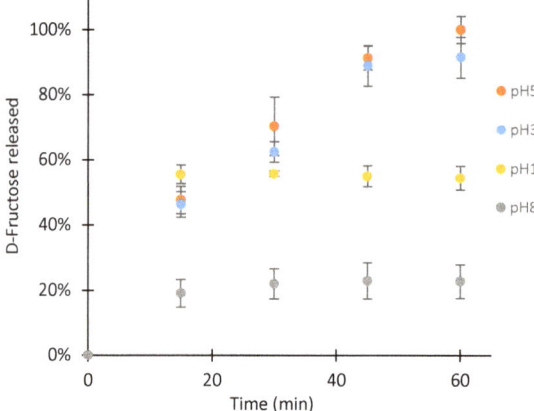

Figure 14. Influence of the pH of the aqueous receiving phase on the D-fructose complex transportation. Extraction process: Tris-HCl 100 mM pH 8.5, $[\text{D-Fru}]_i$ = 100 mM, [3,4-DCPBA] = 100 mM, [Aliquat336®] = 200 mM, MIBK, 343 K. Receiving phase: Citrate buffer 100 mM pH 3 and 5, Tri-HCl pH 8, H2SO4 for pH 1.

2.3. Study of D-Fructose Dehydration in the Receiving Aqueous Phase

D-Fructose dehydration to HMF was carried out in the receiving aqueous phase. We observed previously when studying fructoboronate complex transportation and hydrolysis in a highly acidic aqueous phase the formation of formic and levulinic acids. This confirmed that solutions to minimize HMF degradation by rehydration in the aqueous phase are required. An alternative to strong homogeneous acid has been to use an acidic resin containing strong sulfonic groups. its supported character facilitating a potential regenerability and recyclability of the catalyst. The use of solid acid catalysts prevented the problems of recycling, waste treatment and the risks of use and heating strong acids in liquid form. The type of dehydration catalyst selected for this study was an acidic resin, more particularly a sulfonic resin grafted onto a solid support. Then we have studied the effect of temperature (70 °C, 80 °C and 90 °C) on the yield and the selectivity of the reaction to determine the mass ratio of H^+(resin)/D-fructose (1/1, 2/1 and 3/1). Finally, we have studied the effect of pH on the reaction efficiency. The results are not presented because we have logically observed that the HMF yield increases with the increase of temperature and the H^+/D-fructose ratio. However, until 80 °C no humins are detected but brown colored resin is observed when the temperature is at 90 °C, indicating the formation and adsorption of humins. Therefore, we chose a temperature of 80 °C and a ratio of 3/1 for the following studies (sequential approach).

2.4. Study of Sequential Process

In the formerly optimized conditions, a sequential process was tested. The first step consisted on a simultaneous D-glucose isomerization and D-fructose transportation through a fructoboronate complex, followed by the second step characterized by the hydrolysis of the fructoboronate complex to D-fructose and release of D-fructose in the aqueous receiving phase, and finally by the dehydration of fructose to HMF by heterogeneous catalyst in the third step (Figure 15).

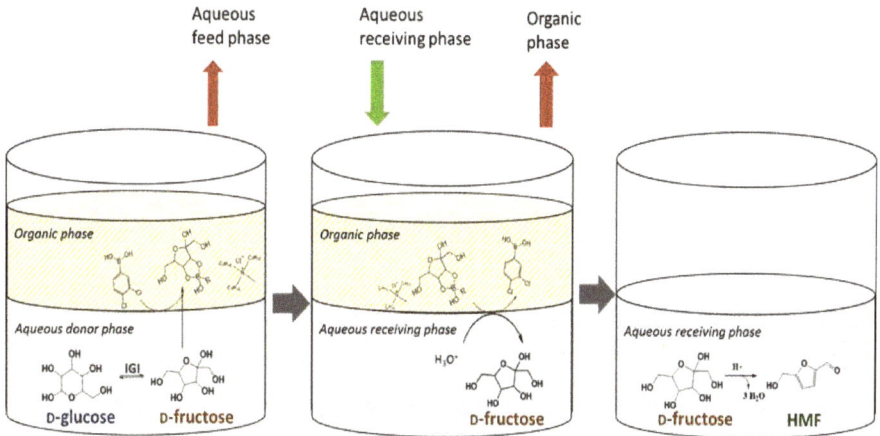

Figure 15. Illustration of the sequential process with isomerization and complexation/transportation step, hydrolysis/release step and dehydration step.

2.4.1. D-Glucose Isomerization and D-Fructose Complexation/Transportation

The aqueous feed phase contained Tris-HCl buffer 100 mM at pH 8.5, D-Glucose and immobilized glucose isomerase at the required amount. The organic layer contained the system determined by the previous study: [3,4-DCPBA] = 100 mM, [Aliquat336®] = 200 mM. The initial aqueous D-glucose concentration was 100 mM. The system was mixed in a reactor where the enzyme was placed in a basket to avoid any contact with the organic layer. Moreover, the use of MIBK, with low solubility in water, as an organic solvent contributed to making the contact between MIBK and enzyme negligible. The system was heated at 343 K for 3 h. The amount of D-glucose and D-fructose were chromatographically determined in the aqueous phase during the process. Table 5 shows the performance of each step. An interesting isomerization yield of 74.5% was obtained showing the actual shifting of the equilibrium of the enzymatic catalysis (55%) by fructose elimination from the media through complexation with the boronic acid. Moreover, according to the extraction yield of D-fructose and D-glucose, 56.5% and 1%, respectively, a very high selectivity of the complexation reaction with the chosen boronic acid towards the D-fructose was actually demonstrated. The partial D-fructose extraction yield of 56.5% showed that the reaction conditions could be still improved.

Table 5. Results obtained from sequential process. The D-Fructose released yield was calculated using the amount of D-Fructose in the organic phase at the end of the 1st transport. The fructose conversion yield was calculated using the amount of fructose in the aqueous receiving phase at the end of the 2nd transport. The HMF total amount was calculated using the initial amount of glucose in the aqueous donor phase.

	Isomerization	1st step		2nd step	3rd step		
		D-Fructose Extracted	D-Glucose Extracted	D-Fructose Released	Fructose Conversion	HMF Yield	HMF Total
%	74.5	56.5	1.56	57.4	52	21.9	5.3
time		3 h		3 h	35 h		41 h

2.4.2. Fructoboronate Complex Hydrolysis, Release of D-Fructose and Dehydration to HMF

The basket of IGI was removed and the aqueous feed phase was discarded and replaced by the receiving aqueous phase with a citrate buffer at 100 mM keeping a pH 3. The system was heated at 343 K for 3 h. In these conditions, the hydrolysis step led to 63% of fructose release in the aqueous phase. Then, after 3 h at 343 K, the organic layer was discarded and the acidic resin was introduced. The system was heated at 363 K. The dehydration step led to a HMF yield of 20% after 10 h. The total HMF yield was 5%.

2.5. Study of Continuous Process

In the best conditions previously determined, a continuous integrated process was tested [33]. The reactor, phases and matter flows are presented in Figure 16. The "coaxial" reactor is composed of two cylinders coaxially placed, a cover enabling introduction of the stirring system and a refrigerant for temperature control. The inner cylinder contains the donor aqueous phase and has a height less than that of the outer cylinder which contains the receiving aqueous phase, which allows the organic phase to be in contact with both aqueous phases. The experiment was conducted with a donor aqueous phase formed by 50 mL of 100 mM Tris-HCl pH 8.5 containing 100 mM of D-glucose, 200 mg IGI, 20 mM $MgSO_4$, 8 mM Na_2SO_3. The receiving aqueous phase was formed by 55 mL of 100 mM sodium citrate buffer, pH 3. The organic phase consisted of 100 mL of MIBK containing 100 mM of 3,4-DCPBA and 200 mM of Aliquat336®. All the phases were maintained at 343 K. The stirring was carried out at 180 rpm, which is the maximum rate to avoid the mixing of the phases, thanks to a stirring blade immersed in the organic phase. This "coaxial" reactor was used to simultaneously perform the isomerization, extraction, hydrolysis and dehydration steps. The monitoring of the D-fructose concentration in the aqueous phases was carried out, making it possible to calculate the efficiency of the continuous process for the simultaneous isomerization, fructose extraction and dehydration. The results are shown in Table 6.

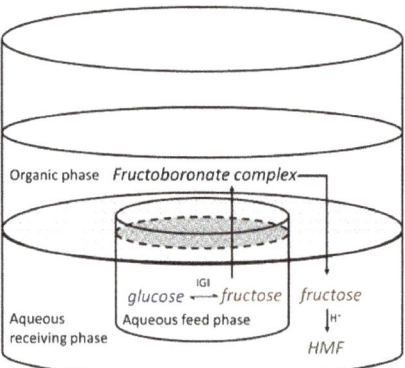

Figure 16. Illustration of the continuous process.

Table 6. Results obtained by continuous process.

	Isomerization	Fructose Extraction	HMF	Selectivity
Yield (%)	70.1	50.2	4.1	70.4

The D-glucose isomerization yield was 70%, with 50% of the formed D-fructose transported and a global HMF yield of 4% (Table 6). Those results are comparable to those obtained by the sequential process. Therefore, this process demonstrates the efficiency of using a cascade chemo-enzymatic catalysis to continuously obtain HMF from D-glucose. However, insufficient agitation of each phase seems to be the main limitation of this reactor configuration. Further improvements of this proof of concept towards enhanced HMF yields and global performances will be reported in due course.

3. Materials and Methods

3.1. Chemicals and Reagents

D-Glucose ($\geq$99.5%), D-Fructose ($\geq$99%), N-methyl-N,N,N-trioctylammonium chloride (Aliquat336®), 3,4-dichlorophenylboronic acid (3,4-DCPBA), 3,5-dichlorophenylboronic acid (3,5-DCPBA) (98%), 2,4-dichlorophenylboronic acid (2,4-DCPBA), 2,3-dichlorophenylboronic acid (2,3-DCPBA), 4-tert-butylphenylboronic acid (4-TBPBA) ($\geq$95%), 4-(trifluoromethyl)phenylboronic acid (4-TFMPBA), Trizma base ($\geq$99.9%), hydrochloric acid (36.8–38%), sulfuric acid (95–97%), Sweetzyme IT®Extra (IGI) ($\geq$350 U/g), Dowex monosphere 650C, levulinic acid (LA) (98%), formic acid (FA) (95%), trisodium citrate ($\geq$99%), 5-hydroxymethylfurfural (HMF) (99%) were purchased from Sigma Aldrich Co. (St. Louis, MO, USA). Phenylboronic acid (PBA) (95%) and 4-methyl-2-pentanone (MIBK) ($\geq$99%) were purchased from Sigma-Fluka (Sigma-Aldrich, St. Louis, MO, USA). Citric acid ($\geq$99%) was purchased from Merck (Merck KGaA, Darmstadt, Germany). Distilled-deionized water (Mili-Q) grad was used whenever necessary, obtained from a MilliQ water purification system (Millipore, Molsheim, France).

3.2. Statistical Analysis

The experiments were performed in triplicate each time. The standard deviation for each experimental result was calculated using Microsoft Excel and reported. The standard deviation for each value was $\leq$5%.

3.3. Characterization

In the conversion/complexation experiments, the determination of reactants, intermediates and products quantities was carried out using a Shimadzu high-performance liquid chromatography (HPLC) (Shimadzu Europa, Duisburg, Germany) instrument equipped with a LC-20ADXR pump, a DGU-20A5R degasser, a SIL-20ACXR autosampler, a SPD-MD20A diode array detector, a CTO-20AC oven and a CBM-20A communicator module piloted by a LabSolution software (LC Driver Ver1.0, Shimadzu Europa, Duisburg, Germany). The column was an Aminex HPX-87H (300 mm × 7.8 mm, 9 μm; Bio-Rad, Hercules, CA, USA). The mobile phase was 5 mM H_2SO_4, using isocratic mode with 0.4 mL.min-1. The chosen wavelength for HMF and levulinic acid detection were 284 nm and 266 nm, respectively. The chosen wavelength for detection of formic acid, D-Glucose and D-Fructose was 195 nm. In the case of monosaccharides, a refractive index detector was also used.

3.4. Aqueous Phase D-Glucose Isomerization (Aqueous Feed Phase) and D-Fructose Complexation/Transportation

3.4.1. D-Glucose Isomerization in Aqueous Feed Phase

We used one of the commercial enzymes most commonly used for the conversion of starches to high fructose syrup, namely immobilized glucose isomerase (IGI, Sweetzyme®IT extra) [31]. D-glucose solution (1 mL) was prepared by mixing D-glucose (100 mM), Na_2SO_3 (8 mM) and $MgSO_4$ (20 mM) in a buffer solution (100 mM) at a selected pH (Tris-HCl buffer for pH 7, 7.5, 8, 8.5, 9, phosphate buffer for pH 6.5, or citrate buffer for pH 4.5) and then introduced in an Eppendorf tube. The solution was incubated at the desired temperature (323 to 363 K) in a Mixing Block (BIOER-102 Thermocell) at 900 rpm. After rehydration in water for 12 h, IGI (10 mg) was put in contact with the aforementioned prepared D-glucose solution. The reaction time was set at 90 min.

3.4.2. D-Fructose Transport in the Organic Phase

First, a D-fructose/D-glucose (25 to 600 mM) solution was prepared in Tris-HCl buffer (100 mM). A volume (600 µL) of this solution was introduced in an Eppendorf tube (solution 1). Aliquat336® (25–300 mM), and arylboronic acid (25–300 mM) were dissolved in MIBK as a solvent to prepare solution 2. Then, 600 µL of the solution 2 were introduced in a second Eppendorf tube. Solutions 1 and 2 were incubated at the desired temperature in a mixing block (BIOER-102 Thermocell) at 750 rpm. After 5 min, both solutions were mixed in an Eppendorf tube placed in the Mixing Block in the same conditions. After D-fructose transport, the aqueous and organic phases were separated by pipetting.

3.4.3. Aqueous Phase D-Fructose Complex Hydrolysis in Receiving Aqueous Phase

Citrate buffer (600 µL) were introduced in an Eppendorf tube and incubated at the required temperature (343 K) in a mixing block at 750 rpm. After 5 min, 600 µL of the organic phase obtained after the extraction procedure and containing the fructoboronate complex were introduced in the tube. After a reaction time of 45 min, the aqueous and organic phases were separated by pipetting.

3.4.4. Study of Sequential Process from Glucose Isomerization to Fructose Dehydration

A 250 mL vessel (Reactor-ReadyTM Lab Reactor, Radley, Figure 15) equipped with a temperature control system (Ministat, Huber temperature control system, HUBER, Chippenham, UK) was used for these experiments.

- Simultaneous D-glucose isomerization by enzymatic catalysis in the aqueous feed phase and D-fructose complexation/transportation in the organic phase.

IGI (0.5 g) was loaded in a stirring basket introduced in a first reactor (Figure 15—left) with 100 mL of a aqueous solution containing D-Glucose (100 mM), Na_2SO_4 (8 mM), $MgCl_2$ (20 mM), in Tris-HCl buffer (100 mM, pH 8.5) at 343 K. Then, 100 mL of MIBK containing 3,4-DCPBA (100 mM) and Aliquat336® (200 mM) were introduced in the reactor. The mixture was stirred via the stirring basket at 200 rpm, with the temperature kept at 343 K for 180 min, before discarding the aqueous feed phase.

- D-fructose complex hydrolysis in the aqueous receiving phase.

The organic phase (100 mL) issued from the first reactor and containing the fructoboronate complex were transferred in a second reactor (Figure 15—middle) and heated at 343 K. Then, 100 mL of a citrate buffer solution (100 mM, pH 3) were introduced. The mixture was stirred with a turbine stirring shaft at 200 rpm and set up at 343 K for 180 min.

4. Conclusions

In this work, we have first provided the conditions of the key parameters (temperature, pH, arylboronic acid structure and concentrations, Aliquat336® concentrations) applied for a sequential process from glucose to HMF through fructose transportation in an organic liquid membrane. The best conditions have been then applied to an unprecedented integrated process in a specifically designed reactor. Therein, enzymatic glucose isomerization and fructose dehydration to HMF, involving an intermediate fructose transportation by complexation with an aryboronic acid-Aliquat336® complex, could be carried out simultaneously in independent aqueous phases. These phases were both in contact with a single organic phase enabling the transportation of fructose from the first aqueous phase to the second. In this process, we thus succeeded in simultaneously enzymatically isomerizing glucose to fructose and dehydrating fructose to HMF by minimizing the number of reaction media and separation steps. Indeed, no intermediate product or phase isolation was necessary. The use of numerous organic solvents or ionic liquid, as previously described in the literature, is limited to a unique organic phase used for fructose transportation. Improvement of performances in terms of HMF yields implying better diffusion of species between the different phases by improving the conception of new reactor systems will be reported in due course.

Author Contributions: Conceptualization, Alexandra Gimbernat and Marie Guehl; Methodology, Nicolas Lopes Ferreira and Egon Heuson; Validation, Rénato Froidevaux, Jean-Sébastien Girardon and Damien Delcroix; Formal Analysis, Mickael Capron and Pascal Dhulster; Investigation, Alexandra Gimbernat and Marie Guehl; Data Curation, Franck Dumeignil; Writing-Original Draft Preparation, Alexandra Gimbernat and Marie Guehl; Supervision, Rénato Froidevaux, Jean-Sébastien Girardon and Damien Delcroix.

Funding: This research received no external funding.

Acknowledgments: This research was funded by IFP Energies nouvelles for funding these studies. Chevreul Institute (FR 2638), Ministère de l'Enseignement Supérieur et de la Recherche, Région Hauts-de-France (CPER ALIBIOTECH), REALCAT platform ('Future Investments' program (PIA), with the contractual reference 'ANR-11-EQPX-0037') and FEDER are acknowledged for supporting and funding partially this work.

Conflicts of Interest: The authors declare no conflict of interest.

References

1. Nanda, S.; Mohammad, J.; Reddy, S.N.; Kozinski, J.A.; Dalai, A.K. Pathways of lignocellulosic biomass conversion to renewable fuels. *Biomass Convers. Biorefin.* **2014**, *4*, 157–191. [CrossRef]
2. Besson, M.; Gallezot, P.; Pinel, C. Conversion of Biomass into Chemicals over Metal Catalysts. *Chem. Rev.* **2014**, *114*, 1827–1870. [CrossRef] [PubMed]
3. Aresta, M.; Dibenedetto, A.; Dumeignil, F. *Biorefinery: From Biomass to Chemicals and Fuels*, 3rd ed.; De Gruyter: Berlin, Germany, 2012; ISBN 978-3-11-026023-6.
4. Agbor, V.B.; Cicek, N.; Sparling, R.; Berlin, A.; Levin, D.B. Biomass pretreatment: Fundamentals toward application. *Biotechnol. Adv.* **2011**, *29*, 675–682. [CrossRef] [PubMed]
5. Bozell, J.J.; Petersen, G.R. Technology development for the production of biobased products from biorefinery carbohydrates—The USA Department of Energy's "Top 10" revisited. *Green Chem.* **2010**, *12*, 539–545. [CrossRef]
6. Van Heerden, C.D.; Nicol, W. Continuous succinic acid fermentation by *Actinobacillus succinogenes*. *Biochem. Eng. J.* **2013**, *73*, 5–11. [CrossRef]
7. Girard, E.; Delcroix, D.; Cabiac, A. Catalytic conversion of cellulose to C_2–C_3 glycols by dual association of a homogeneous metallic salt and a perovskite-supported platinum catalyst. *Catal. Sci.* **2016**, *6*, 5534–5542. [CrossRef]
8. Shanks, B.H. Unleashing Biocatalysis/Chemical Catalysis Synergies for Efficient Biomass Conversion. *ACS Chem. Biol.* **2007**, *2*, 533–535. [CrossRef] [PubMed]
9. Vennestrøm, P.N.R.; Taarning, E.; Christensen, C.H.; Pedersen, S.; Grunwaldt, J.D.; Woodley, J.M. Chemoenzymatic Combination of Glucose Oxidase with Titanium Silicalite-1. *ChemCatChem* **2010**, *2*, 943–945. [CrossRef]
10. Denard, C.A.; Hartwig, J.F.; Zhao, H. Multistep One-Pot Reactions Combining Biocatalysts and Chemical Catalysts for Asymmetric Synthesis. *ACS Catal.* **2013**, *3*, 2856–2864. [CrossRef]
11. Dumeignil, F. Public Serv. *Rev. Eur. Union* **2011**, *22*, 528.
12. Dumeignil, F. Chemical Catalysis and Biotechnology: From a Sequential Engagement to a One-Pot Wedding. *Chem. Ing. Tech.* **2014**, *86*, 1496–1508. [CrossRef]
13. Van Putten, R.; Van der Waal, J.C.; de Jong, E.; Rasrendra, C.B.; Heeres, H.G.; de Vries, J.G. Hydroxymethylfurfural, A Versatile Platform Chemical Made from Renewable Resources. *Chem. Rev.* **2013**, *113*, 1499–1597. [CrossRef] [PubMed]
14. Moliner, M.; Román-leshkov, Y.; Davis, M.E. Tin-containing zeolites are highly active catalysts for the isomerization of glucose in water. *Proc. Natl. Acad. Sci. USA* **2010**, *107*, 6164–6168. [CrossRef] [PubMed]
15. Pacheco, J.J.; Davis, M.E. Synthesis of terephthalic acid via Diels-Alder reactions with ethylene and oxidized variants of 5-hydroxymethylfurfural. *Proc. Natl. Acad. Sci. USA* **2014**, *111*, 8363–8367. [CrossRef] [PubMed]
16. Lin, Z.; Nikolakis, V.; Ierapetritou, M. Alternative Approaches for p-Xylene Production from Starch: Techno-Economic Analysis. *Ind. Eng. Chem. Res.* **2014**, *53*, 10688–10699. [CrossRef]
17. De Jong, E.; Dam, M.A.; Sipos, L.; Gruter, G. Furandicarboxylic Acid (FDCA), a Versatile Building Block for a Very Interesting Class of Polyesters. *Am. Chem. Soc.* **2012**, *1*, 1–13.

18. Boisen, A.; Christensen, T.B.; Fu, W.; Gorbanev, Y.Y.; Hansen, T.S.; Jensen, J.S.; Klitgaard, S.K.; Pedersen, S.; Riisager, A.; Stahlberg, T.; et al. Process Integration for the Conversion of Glucose to 2,5-Furandicarboxylic Acid. *Chem. Eng. Res. Des.* **2009**, *87*, 1318–1327. [CrossRef]
19. Akien, G.R.; Qi, L.; Horvath, I.T. Molecular mapping of the acid catalysed dehydration of fructose. *Chem. Commun.* **2012**, *48*, 5850–5852. [CrossRef] [PubMed]
20. Parker, K.; Salas, M.; Nwosu, V.C. High fructose corn syrup: Production, uses and public health concerns *Biotechnol. Mol. Biol. Rev.* **2010**, *5*, 71–78.
21. Li, H.; Yang, S.; Saravanamurugan, S.; Riisager, A. Glucose Isomerization by Enzymes and Chemo-catalysts: Status and Current Advances. *ACS Catal.* **2017**, *7*, 3010–3029. [CrossRef]
22. Delidovich, I.; Palkovits, R. Catalytic Isomerization of Biomass-Derived Aldoses: A Review. *ChemSusChem* **2016**, *9*, 547–561. [CrossRef] [PubMed]
23. Choudhary, V.; Pinar, A.B.; Lobo, R.F.; Vlachos, D.G.; Sandler, S.I. Comparison of Homogeneous and Heterogeneous Catalysts for Glucose-to-Fructose Isomerization in Aqueous Media. *ChemSusChem* **2013**, *6*, 2369–2376. [CrossRef] [PubMed]
24. Marianou, A.A.; Michailof, C.M.; Pineda, A.; Iliopoulou, E.F.; Triantafyllidis, K.S.; Lappas, A.A. Glucose to Fructose Isomerization in Aqueous Media over Homogeneous and Heterogeneous Catalysts. *ChemCatChem* **2016**, *8*, 1100–1110. [CrossRef]
25. Zhao, S.; Guo, X.; Bai, P.; Lv, L. Catalytic Isomerization of Biomass-Derived Aldoses: A Review. *Asian J. Chem.* **2014**, *26*, 4537–4542.
26. Demerdash, M.; Attia, R.M. Equilibrium kinetics of D-glucose to D-fructose isomerization catalyzed by glucose isomerase enzyme from *Streptomyces phaeochromogenus*. *Zentralbl. Mikrobiol.* **1992**, *147*, 297–303. [CrossRef]
27. Gaily, M.H.; Elhassan, B.M.; Abasaeed, A.E.; Al-shrhan, M. Isomerization and Kinetics of Glucose into Fructose. *Int. J. Eng. Technol.* **2010**, *10*, 1–6.
28. Takasaki, Y. Kinetic and Equilibrium Studies on D-Glucose-D-Fructose Isomerization Catalyzed by Glucose Isomerase from *Streptomyces* sp. *Agric. Biol. Chem.* **1967**, *31*, 309–313. [CrossRef]
29. Huang, H.; Denard, C.A.; Alamillo, R.; Crisci, A.J.; Miao, Y.; Dumesic, J.A.; Scott, S.L.; Zhao, H. Tandem Catalytic Conversion of Glucose to 5-hydroxymethylfurfural with an Immobilized Enzyme and a Solid Acid. *ACS Catal.* **2014**, *4*, 2165–2172. [CrossRef]
30. Huang, R.; Qi, W.; Su, R.; He, Z. Integrating Enzymatic and Acid Catalysis to Convert Glucose into 5-Hydroxymethylfurfural. *Chem. Commun. (Camb.)* **2010**, *46*, 1115–1125. [CrossRef] [PubMed]
31. Delidovich, I.; Palkovits, R. Fructose production via extraction-assisted isomerization of glucose catalyzed by phosphates. *Green Chem.* **2016**, *18*, 5822–5830. [CrossRef]
32. Alipour, S.; Relue, P.A.; Viamajala, S.; Varanasi, S. High yield 5-(hydroxylmethyl)furfural Production from Biomass Sugars under Facile Reaction Conditions: A Hybrid Enzyme- and Chemo-Catalytic Technology. *Green Chem.* **2016**, *18*, 4990–5002. [CrossRef]
33. Gimbernat, A.; Guehl, M.; Capron, M.; Lopes-Ferreira, N.; Froidevaux, R.; Girardon, J.S.; Dhulster, P.; Delcroix, D.; Dumeignil, F. Hybrid Catalysis: A Suitable Concept for the Valorization of Biosourced Saccharides to Value-Added Chemicals. *ChemCatChem* **2017**, *9*, 2080–2888. [CrossRef]
34. Bhosale, S.H.; Rao, M.B.; Deshpande, V.V. Molecular and Industrial Aspects of Glucose Isomerase. *Microbiol. Rev.* **1996**, *60*, 280–290. [PubMed]
35. McKay, G.; Tavlarides, T. Enzymatic isomerization kinetics of D-Glucose to D-Fructose. *J. Mol. Catal.* **1979**, *6*, 57–65. [CrossRef]
36. Morin, G.T.; Paugam, M.F.; Hughes, M.P.; Smith, B.D. Boronic Acids Mediate Glycoside Transport through a Liquid Organic Membrane via Reversible Formation of Trigonal Boronate Esters. *J. Org. Chem.* **1994**, *59*, 2724–2730. [CrossRef]
37. Karpa, M.J.; Duggan, P.J.; Griffin, G.J.; Freudigmann, S.J. Competitive Transport of Reducing Sugars Through a Lipophilic Membrane Facilitated by Aryl Boron Acids. *Tetrahedron* **1997**, *53*, 3669–3676. [CrossRef]
38. Westmark, P.R.; Gardiner, S.J.; Smith, B.D. Selective Monosaccharide Transport through Lipid Bilayers using Boronic Acid Carriers. *J. Am. Chem. Soc.* **1996**, *118*, 11093–11100. [CrossRef]

39. Shinbo, T.; Nishimura, K.; Yamaguchi, T.; Sugiura, M. Uphill Transport of Monosaccharides across an Organic Liquid Membrane. *J. Chem. Soc. Chem. Commun.* **1986**, 349–358. [CrossRef]
40. Paugam, M.; Riggs, J.A.; Smith, B.D. High fructose syrup production using fructose-selective liquid membranes. *ChemComm* **1996**, *22*, 2539–2540. [CrossRef]

 © 2018 by the authors. Licensee MDPI, Basel, Switzerland. This article is an open access article distributed under the terms and conditions of the Creative Commons Attribution (CC BY) license (http://creativecommons.org/licenses/by/4.0/).

Article

Selective Reduction of Ketones and Aldehydes in Continuous-Flow Microreactor—Kinetic Studies

Katarzyna Maresz [1],*, Agnieszka Ciemięga [1] and Julita Mrowiec-Białoń [1,2]

1. Institute of Chemical Engineering Polish Academy of Sciences, Bałtycka 5, 44-100 Gliwice, Poland; akoreniuk@iich.gliwice.pl (A.C.); jmrowiec@polsl.pl (J.M.-B.)
2. Department of Chemical Engineering and Process Design, Silesian University of Technology, Ks. M. Strzody 7, 44-100 Gliwice, Poland
* Correspondence: k.kisz@iich.gliwice.pl; Tel.: +48-32-231-08-11

Received: 24 April 2018; Accepted: 18 May 2018; Published: 22 May 2018

Abstract: In this work, the kinetics of Meerwein–Ponndorf–Verley chemoselective reduction of carbonyl compounds was studied in monolithic continuous-flow microreactors. To the best of our knowledge, this is the first report on the MPV reaction kinetics performed in a flow process. The microreactors are a very attractive alternative to the batch reactors conventionally used in this process. The proposed micro-flow system for synthesis of unsaturated secondary alcohols proved to be very efficient and easily controlled. The microreactors had reactive cores made of zirconium-functionalized silica monoliths of excellent catalytic properties and flow characteristics. The catalytic experiments were carried out with the use of 2-butanol as a hydrogen donor. Herein, we present the kinetic parameters of cyclohexanone reduction in a flow reactor and data on the reaction rate for several important ketones and aldehydes. The lack of diffusion constraints in the microreactors was demonstrated. Our results were compared with those from other authors and demonstrate the great potential of microreactor applications in fine chemical and complex intermediate manufacturing.

Keywords: Meerwein–Ponndorf–Verley reduction; kinetics; flow microreactor

1. Introduction

A reduction of carbonyl bond is a widespread route for the synthesis of alcohols. However, the reaction, classically catalyzed by noble metals and carried out in the presence of molecular hydrogen, reveals significant limitations, including low selectivity, high sensitivity to sulfur-containing substrates, and high-pressure requirements. The pharmaceutical industry is concerned with the purity of its products. The Meerwein–Ponndorf–Verley (MPV) reaction is an attractive method of synthesizing unsaturated alcohols from ketones or aldehydes using secondary alcohols instead of gaseous hydrogen. According to a generally accepted mechanism of the MPV reaction, the carbonyl group acts as a hydrogen acceptor and alcohol as a hydrogen source. The hydrogen transfer occurs when both substrates are simultaneously coordinated to the same Lewis acidic site (Scheme 1). The formation of a six-membered transition state ring is considered to be a determining step in the reaction rate.

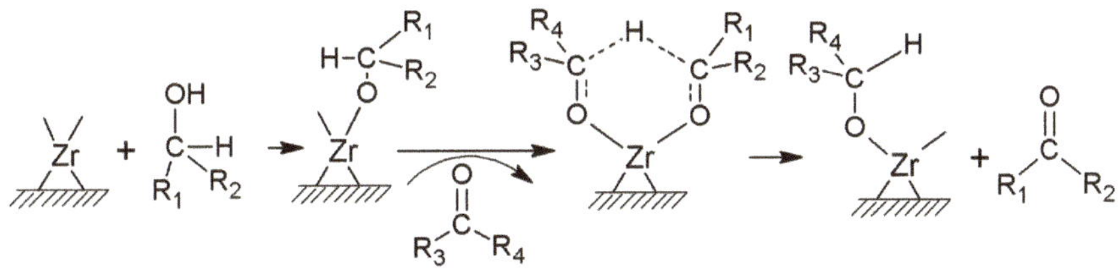

Scheme 1. The mechanism of Meerwein–Ponndorf–Verley (MPV) reduction.

Inexpensive and non-toxic hydrogen donors and catalysts, mild reaction conditions, and exceptional chemoselectivity render this method of reduction favorable over alternatives. Among many active species, such as Zr [1,2], Al [3,4], Mg [3,5], and B [6,7], which are considered to be active catalysts for MPV reduction, zirconium has been shown to be one of the most promising. In the literature, batch processes are predominantly described with the use of numerous catalysts, e.g., homogeneous alkoxides [8,9], zeolites [10,11], mesoporous sieves [12–14], and hydrotalcite [15]. Nevertheless, the tedious separation of homogenous catalysts at the end of the process leads to its deactivation and non-reusability. Powdered catalysts ensure significant benefits over its homogeneous counterparts. However, filtration is an additional time- and cost-consuming step in the technological line. Flow microreactors allow one to overcome these drawbacks and have additional advantages, i.e., high surface-to-volume ratio, improved reaction parameter control, a small equipment size, and a flexibility of module arrangement.

Flow chemistry is perspective and still not explored field in the area of chemoselective MPV reduction. Battilocchio et al. reported the protocol for synthesizing various alcohols from aromatic and aliphatic carbonyl compounds using a packed-bed reactor filled with zirconium hydroxide catalyst [16]. In our previous works [17–19], we demonstrated excellent activity of zirconium-doped silica monolithic microreactors in cyclohexanone reductions and their improved performance compared with the batch process. It was shown that zirconium species terminated with propoxy ligands featured the highest activity in MPV reduction among various Lewis centers immobilized onto monoliths' surfaces used as reactive cores in microreactors. Extensive studies of structural, physicochemical, and catalytic properties revealed the high efficiency of the proposed microreactors.

In this work, we present the kinetic studies of MPV reduction with the use of various carbonyl compounds and 2-butanol as a hydrogen donor. The experiments were performed in continuous-flow zirconium-propoxide-functionalized microreactors of different lengths.

We determined the kinetic parameters, hardly presented for the MPV reduction process carried out in a flow regime. The results of the flow and batch reactors were compared with those of other authors. The kinetic data are crucial to determine the optimum process conditions through the selection of appropriate catalysts and reaction parameters. The knowledge of basic issues related to the course of reactions allows one to set new, more effective paths for conducting processes. Despite the possibility of theoretical computer simulation of the behavior of the reaction system, the experimental determination of the kinetic equation parameters is still necessary for the development of the reactor model.

2. Results

The characterization of siliceous monolithic microreactors functionalized with zirconium species and their catalytic properties were described in our previous papers [17,18]. Kinetic studies were performed in the microreactors featured by the exceptionally low back-pressure. It resulted from the unique structure of cylindrical-shape monoliths applied as reactive cores [20]. Briefly, monoliths were characterized by a continuous, macroporous flow-through structure (macropores in the range of 30–50 µm), extended with mesopores of bimodal size distributions (3/20 nm) and a high surface area (340 $m^2 \cdot g^{-1}$) (Figure 1). The monoliths were functionalized with zirconium propoxide to obtained

Lewis acid sites on the silica surface. The zirconium cations were terminated by propoxy/hydroxo ligands, which appeared to be very active catalytic centers [18,19]. The concentration of zirconium, determined by ICP MS analysis, was 7.03 wt % (in relation to the mass of silica support) in all studied microreactors.

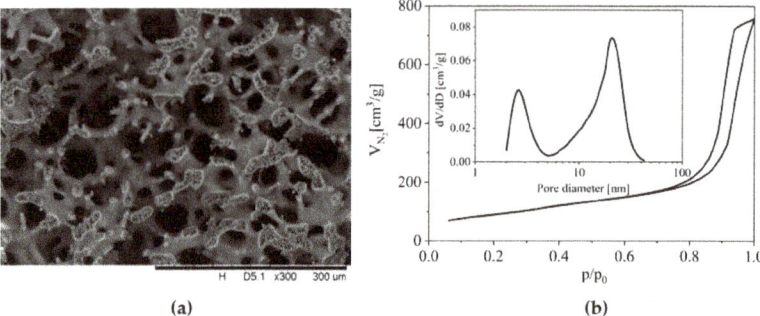

Figure 1. (a) SEM image of silica monolith structure; (b) nitrogen adsorption/desorption isotherm and pore size distribution (insert).

Material functionalization with zirconium precursor did not considerably influence the structural properties of the support. Only a small decline of the surface area and mesopore volume was observed. All features have been preserved after multiple reaction cycles.

Detailed studies of the kinetic experiment were performed for the MPV reduction of cyclohexanone with 2-butanol. First, we checked whether diffusion or activation controls the reaction rate. External mass transfer limitations are a common phenomenon in porous materials. To exclude the impact of diffusional effects on the reaction kinetics, the performance of microreactors with monolithic cores of different lengths were compared with respect to the same residence time, equal to 5 min. Similar conversions of cyclohexanone, about 40%, were achieved in all microreactors of a 1–8 cm length, which evidenced the lack of transport constraints (Figure 2).

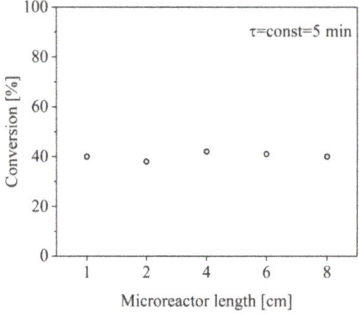

Figure 2. Conversion of cyclohexanone in microreactors of different lengths for a residence time of 5 min.

The results of kinetic experiments for cyclohexanone reduction are depicted in Figure 3. Figure 3a shows the single experimental run carried out for 6 h at 95 °C in a 6 cm long microreactor. The reaction rate constant (Figure 3c) was determined by assuming the first order kinetics and was calculated using Equation (1):

$$\ln(C_0/C) = k\tau \qquad (1)$$

where C_0 [mmol·cm^{-3}] is the initial concentration of ketone, C [mmol·cm^{-3}] the substrate concentration after the reaction in the microreactor of a fixed length, k [min^{-1}] the rate constant, τ [min] the residence time.

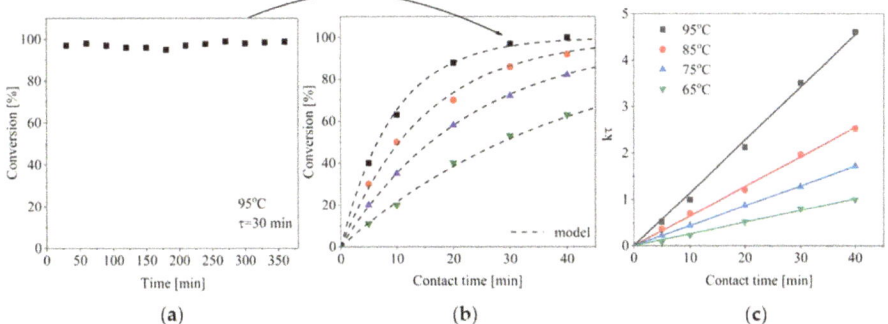

Figure 3. (a) Conversion of cyclohexanone in the 6-cm-long microreactor. (b) Catalytic results for MPV reduction at 65 °C (▼), 75 °C (▲), 85 °C (●), and 95 °C (■) in the microreactor (points—experiments, lines-model). (c) Plot of ln (C_0/C) versus contact time.

Figure 3b confirms good agreement between the experimental data and the model prediction of the first-order kinetics for the MPV process carried out at different temperatures and contact times in the range of 5–40 min. Each point corresponds to the average conversion obtained in microreactor during 6-h-long tests.

The linear relationship rate constant vs. the contact time was observed through the whole temperature range used in the experiments. The experimental data are in line with those calculated from first-order kinetics equation.

Temperature dependence of the rate constant was examined in the range of 65–95 °C and is shown in Figure 4. The activation energy was estimated from linear regression analysis, and it appeared to be 52 kJ·mol^{-1}, and frequency factor k∞ = 2.69 min^{-1}.

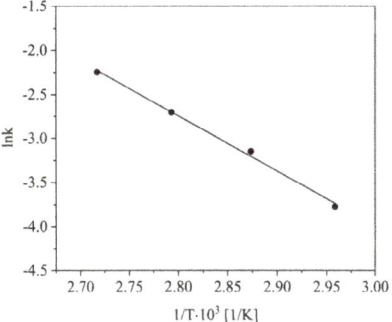

Figure 4. Arrhenius plot for the MPV reduction of cyclohexanone.

The studies of the MPV reduction in the flow process were extended to other ketones and aldehydes, and the results, i.e., the conversion, productivity, and reaction rate constant are summarized in Table 1. Analysis of these data shows that aldehydes are easier to be reduced than ketones. The reaction rate constant of benzaldehyde reduction was 0.212 min^{-1}, while that for cyclohexanone was twice lower. Battilocchio et al. also found that aldehydes, compared with aliphatic ketones, required a shorter reaction time for complete conversion to alcohols [16]. Steric hindrance is a prevalent factor that affects the reactivity of the substrates. Citral is a mixture of cis- and trans-isomers signified as neral and geranial, respectively. The MPV reduction of citral reveals the preferential formation of its trans product. The yield of geraniol was 20% higher than nerol. The orientation of the carbonyl group to the molecular chain in geranial reduces steric limitations for binding between the Zr center and C=O.

Table 1. Data from catalytic experiments.

Substrate	K^1 [min^{-1}]	Conversion [1,*] [%]	Productivity [1,*] [mmol·g$_{cat}^{-1}$·h^{-1}]	Productivity [2] [mmol·g$_{cat}^{-1}$·h^{-1}]	Productivity [3] [mmol·g$_{cat}^{-1}$·h^{-1}]
cyclohexanone	0.106	88	2.22	-	2.28 [21]
2-cyclohexen-1-one	0.021	35	0.9	-	0.11 [15]
benzaldehyde	0.212	99	2.64	1.14 [16]	0.36 [10]
cinnamaldehyde	0.081	80	1.92	0.66 [16]	0.48 [5]
citral	0.031/0.047 [4]	46/61	0.48/0.9	-	0.54 [5]
acetophenone	0.026	40	0.96	0.48 [16]	0.12 [22]
hexanal	0.08	80	2.1	-	-
decanal	0.041	56	1.32	-	0.72 [23]

[1] this work (reaction temp. 95 °C), [2] in the flow process (data form literature), [3] in the batch process (data form literature); [4] nerol/geraniol; * data for the 4-cm-long microreactor.

The impact of both the steric and electronic effect can be observed in the case of cinnamaldehyde reduction. The low activity can be attributed to the presence of a bulky chain in this substrate and double bond. The productivity in the reduction process of cinnamaldehyde was significantly lower than that for benzaldehyde. The reduction of unsaturated ketone, 2-cyclohexen-1-on, is difficult without affecting the conjugated C=C bond. A significant decrease in conversion was observed compared to the saturated cyclic carbonyl compound (cyclohexanone); nevertheless, the product was obtained with 100% selectivity. Aromatic ketones, compared with cyclic ketones, are more difficult to reduce due to the resonance and inductive effect of the benzene ring [22]. It explains the difference in conversions of cyclohexanone and acetophenone. No products arising from the Tischenko cross reaction or from the aldol condensation [24], were detected for any of the investigated substrates. This evidenced the excellent selectivity of the proposed catalyst.

It should furthermore be highlighted that alcohols produced by the selective hydrogenation of carbonyl-bearing substrates are fine chemicals of primary interest. They are used as flavor additives, and intermediates in drug production. Products of the esterification of cinnamyl alcohol, indole-3-acetic acid, or α-lipoic acid were studied for their antioxidant and anti-inflammatory activity [25], and graniol was checked in colon cancer chemoprevention and treatment [26].

Our results were compared to the literature data for flow and batch processes. To the best of our knowledge, the MPV process using heterogeneous powdered catalyst at flow conditions has only been published in one paper [16]. The productivities of ZrO_2-based reactors toward benzyl alcohol, cinnamyl alcohol, and 1-phenylethanol are nearly 4 times lower than those achieved in the studied monolithic microreactors. Table 1 shows the productivities from batch reactors obtained by other research. They are significantly lower than those obtained in the proposed flow system (except one case). It was due to excellent mixing and mass transfer conditions, offered by innovative monolithic microreactor characterized by high surface-to-volume ratios. Application of the proposed continuous-flow microreactor for the MPV process offers not only enhanced productivity, but also facilitates process handling by excluding contact with reaction media. The latter is of importance when working with dangerous substances.

3. Materials and Methods

Microreactors were fabricated using silica monoliths as cores. The monoliths were obtained by combined sol-gel and phase separation methods described in details in [20]. Briefly, polyethylene glycol (PEG 35 000, Sigma-Aldrich, St. Louis, MO, USA) was dissolved in 1 M nitric acid (Avantor, 65%, Gliwice, Poland). The mixture was cooled in the ice bath and subsequently tetraethoxysilane (TEOS, ABCR, 99%, Karlsruhe, Germany) was added dropwise. After 30 min of stirring, cetyltrimethylammonium bromide (CTAB, Sigma-Aldrich) was added. After complete dissolution, the sol was transferred to the polypropylene molds and stored at 40 °C for 8 days for gelation and aging. Next, rod-shaped monoliths were washed with deionized water and treated with 1 M ammonia solution (Avantor, 25%) for 8 h at 90 °C. Afterward, the materials were washed, dried at 40 °C for 3 days, and finally calcined at 550 °C for 8 h to remove organic templates. The prepared monoliths of diameter 4.5 mm were put into heat-shrinkable tubes and equipped with connectors to obtain flow microreactors.

Zirconium propoxide moieties were grafted onto silica carriers, maintaining Zr/Si ratio fixed at 0.14 and using a solution of zirconium(IV) propoxide (Sigma Aldrich, 70% in 1-propanol) in anhydrous ethanol (Avantor, 99.8%). Reactive cores were impregnated with the solution and kept for 24 h at 70 °C. Finally, they were washed with ethanol and dried at 110 °C in nitrogen flow conditions.

Structural properties were analyzed by electron microscopy (TM 30000, Hitachi, Chiyoda, Tokyo, Japan) and adsorption–desorption measurements (ASAP 2020, Micromeritics, Norcross, GA, USA).

The continuous-flow microreactors were tested in the MPV reduction of various carbonyl compounds (cyclohexanone: Sigma Aldrich, 99%, benzaldehyde: Sigma-Aldrich, 99%, acetophenone: Acros, Geel, Belgium, 98%, cinnamaldehyde: Acros, 99%, citral: Roth, 95–98%, Karlsruhe, Germany, hexanal: Aldrich, Saint Louis, MO, USA, 98%, nonanal: Aldrich, 95%, 2-cyclohexen-1-one: Acros, 97%) with 2-butanol (Avantor 99%). The molar ratio of substrates was 1:52. The flow rate was changed in the range of 0.03–0.24 $cm^3 \cdot min^{-1}$. The progress and selectivity of the reaction were monitored by gas chromatography (FID detector, HP-5 column, 7890A, Agilent, Santa Clara, CA, USA).

The kinetic experiments were carried out in setup shown in Figure 5 for a flow rate fixed at 0.03 $cm^3 \cdot min^{-1}$, at the temperature range of 65–95 °C. The mass of the 1-cm-long microreactor was 0.0375 g. The length of reactive cores was changed from 1 to 8 cm (weight form 0.0375–0.3 g) and enabled the obtainment of data for different residence times, calculated using Equation (2):

$$\tau = \frac{m_k \cdot l \cdot V_T}{F} = \frac{0.0375 \cdot l \cdot V_T}{F} \qquad (2)$$

where m_k is the mass of monolith per length unit [$g \cdot cm^{-1}$], l is the microreactor length [cm], V_T is the total pore volume, equal to −4 [$cm^3 \cdot g^{-1}$] (data from mercury porosimetry), and F is the flow rate [$cm^3 \cdot min^{-1}$].

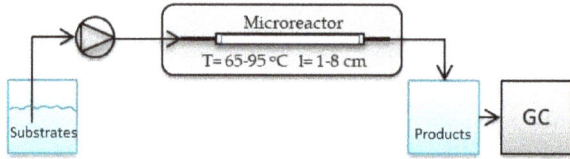

Figure 5. Scheme of the reaction setup.

4. Conclusions

The outstanding potential of continuous-flow microreactors in the area of selective reduction of carbonyl compounds was demonstrated. Broad-range and long-term experiments were conducted to determine the kinetics of the MPV reaction of cyclohexanone in the zirconium-functionalized flow microreactors. The lack of diffusion constraints in the microreactors was shown. The activation energy was calculated to be 52 kJ·mol^{-1}. Moreover, reaction rate constants for several ketones and aldehydes were collected. The rate of the process is necessary to design the apparatus and reaction systems. Significant differences in process efficiency were recorded for various carbonyl compounds. They were assigned to steric effects caused by bulky chains, electronic effects of an additional double bond, and an inductive effect of the benzene ring. We believe that the proposed system can be effectively exploited for fine chemical and pharmaceutical production.

Author Contributions: A.C. and K.M. conceived, designed, and performed the experiments; A.C., K.M., and J.M.-B. analyzed the data and wrote the paper.

Funding: This research was funded by the National Science Centre, Poland: DEC-2017/01/X/ST8/00083.

Acknowledgments: The financial support of the National Science Centre, Poland (project no. DEC-2017/01/X/ST8/00083), is gratefully acknowledged.

Conflicts of Interest: The authors declare no conflict of interest.

References

1. Liu, S.H.; Jaenicke, S.; Chuah, G.K. Hydrous zirconia as a selective catalyst for the Meerwein–Ponndorf–Verley reduction of cinnamaldehyde. *J. Catal.* **2002**, *206*, 321–330. [CrossRef]
2. Plessers, E.; Fu, G.X.; Tan, C.Y.X.; De Vos, D.E.; Roeffaers, M.B.J. Zr-based MOF-808 as Meerwein–Ponndorf–Verley reduction catalyst for challenging carbonyl compounds. *Catalysts* **2016**, *6*, 104. [CrossRef]
3. Axpuac, S.; Aramendía, M.A.; Hidalgo-Carrillo, J.; Marinas, A.; Marinas, J.M.; Montes-Jiménez, V.; Urbano, F.J.; Borau, V. Study of structure–performance relationships in Meerwein–Ponndorf–Verley reduction of crotonaldehyde on several magnesium and zirconium-based systems. *Catal. Today* **2012**, *187*, 183–190. [CrossRef]
4. Zapilko, C.; Liang, Y.C.; Nerdal, W.; Anwander, R. A general strategy for the rational design of size-selective mesoporous catalysts. *Chem. Eur. J.* **2007**, *13*, 3169–3176. [CrossRef] [PubMed]
5. Aramendía, M.A.A.; Borau, V.; Jiménez, C.; Marinas, J.M.; Ruiz, J.R.; Urbano, F. Reduction of α,β-unsaturated aldehydes with basic MgO/M$_2$O$_3$ catalysts (M=Al, Ga, In). *Appl. Catal. A Gen.* **2003**, *249*, 1–9. [CrossRef]
6. Uysal, B.; Oksal, B.S. A new method for the chemoselective reduction of aldehydes and ketones using boron tri-isopropoxide, B(O^{i}Pr)$_3$: Comparison with boron tri-ethoxide, B(OEt)$_3$. *J. Chem. Sci.* **2011**, *123*, 681–685. [CrossRef]
7. Uysal, B.; Oksal, B.S. New heterogeneous B(OEt)$_3$-MCM-41 catalyst for preparation of α,β-unsaturated alcohols. *Res. Chem. Intermed.* **2015**, *41*, 3893–3911. [CrossRef]
8. Meerwein, H.; Schmidt, R. Ein neues verfahren zur reduktion von aldehyden und ketonen. *Justus Liebigs Ann. Chem.* **1925**, *444*, 221–238. [CrossRef]
9. Ooi, T.; Miura, T.; Takaya, K.; Ichikawa, H.; Maruoka, K. Zr(OBut)$_4$ as an effective promoter for the Meerwein–Ponndorf–Verley alkynylation and cyanation of aldehydes: Development of new asymmetric cyanohydrin synthesis. *Tetrahedron* **2001**, *57*, 867–873. [CrossRef]

10. Wang, J.; Okumura, K.; Jaenicke, S.; Chuah, G.K. Post-synthesized zirconium-containing beta zeolite in Meerwein–Ponndorf–Verley reduction: Pros and cons. *Appl. Catal. A Gen.* **2015**, *493*, 112–120. [CrossRef]
11. Creyghton, E.J.; Downing, R.S. Shape-selective hydrogenation and hydrogen transfer reactions over zeolite catalysts. *J. Mol. Catal. A Chem.* **1998**, *134*, 47–61. [CrossRef]
12. Zhang, B.; Xie, F.; Yuan, J.; Wang, L.; Deng, B.X. Meerwein–Ponndorf–Verley reaction of acetophenone over ZrO_2-La_2O_3/MCM-41: Influence of loading order of ZrO_2 and La_2O_3. *Catal. Commun.* **2017**, *92*, 46–50. [CrossRef]
13. Sushkevich, V.L.; Ivanova, I.I.; Tolborg, S.; Taarning, E. Meerwein–Ponndorf–Verley-Oppenauer reaction of crotonaldehyde with ethanol over Zr-containing catalysts. *J. Catal.* **2014**, *316*, 121–129. [CrossRef]
14. Rodriguez-Castellon, E.; Jimenez-Lopez, A.; Maireles-Torres, P.; Jones, D.I.; Roziere, J.; Trombetta, M.; Busca, G.; Lenarda, M.; Storaro, L. Textural and structural properties and surface acidity characterization of mesoporous silica-zirconia molecular sieves. *J. Solid State Chem.* **2003**, *175*, 159–169. [CrossRef]
15. Jimenez-Sanchidrian, C.; Ruiz, J.R. Tin-containing hydrotalcite-like compounds as catalysts for the Meerwein–Ponndorf–Verley reaction. *Appl. Catal. A Gen.* **2014**, *469*, 367–372. [CrossRef]
16. Battilocchio, C.; Hawkins, J.M.; Ley, S.V. A mild and efficient flow procedure for the transfer hydrogenation of ketones and aldehydes using hydrous zirconia. *Org. Lett.* **2013**, *15*, 2278–2281. [CrossRef] [PubMed]
17. Koreniuk, A.; Maresz, K.; Mrowiec-Białoń, J. Supported zirconium-based continuous-flow microreactor for effective Meerwein–Ponndorf–Verley reduction of cyclohexanone. *Catal. Commun.* **2015**, *64*, 48–51. [CrossRef]
18. Ciemięga, A.; Maresz, K.; Mrowiec-Białoń, J. Continuous-flow chemoselective reduction of cyclohexanone in a monolithic silica-supported $Zr(OPr^i)_4$ multichannel microreactor. *Microporous Mesoporous Mater.* **2017**, *252*, 140–145. [CrossRef]
19. Maresz, K.; Ciemięga, A.; Mrowiec-Białoń, J. Meervein–Ponndorf–Vereley reduction of carbonyl compounds in monolithic siliceous microreactors doped with Lewis acid centers. *Appl. Catal. A Gen.* **2018**, *560*, 111–118.
20. Ciemięga, A.; Maresz, K.; Malinowski, J.J.; Mrowiec-Białoń, J. Continuous-flow monolithic silica microreactors with arenesulfonic acid groups: Structure-catalytic activity relationships. *Catalysts* **2017**, *7*, 255. [CrossRef]
21. Li, G.; Fu, W.H.; Wang, Y.M. Meervvein–Ponndorf–Verley reduction of cyclohexanone catalyzed by partially crystalline zirconosilicate. *Catal. Commun.* **2015**, *62*, 10–13. [CrossRef]
22. Corma, A.; Domine, M.E.; Valencia, S. Water-resistant solid Lewis acid catalysts: Meerwein–Ponndorf–Verley and Oppenauer reactions catalyzed by tin-beta zeolite. *J. Catal.* **2003**, *215*, 294–304. [CrossRef]
23. Ruiz, J.R.; Jiménez-Sanchidrián, C.; Hidalgo, J.M.; Marinas, J.M. Reduction of ketones and aldehydes to alcohols with magnesium–aluminium mixed oxide and 2-propanol. *J. Mol. Catal. A Chem.* **2006**, *246*, 190–194. [CrossRef]
24. Aramendia, M.A.; Borau, V.; Jimenez, C.; Marinas, J.M.; Ruiz, J.R.; Urbano, F.J. Activity of basic catalysts in the Meerwein–Ponndorf–Verley reaction of benzaldehyde with ethanol. *J. Colloid Interface Sci.* **2001**, *238*, 385–389. [CrossRef] [PubMed]
25. Theodosis-Nobelos, P.; Kourti, M.; Tziona, P.; Kourounakis, P.N.; Rekka, E.A. Esters of some non-steroidal anti-inflammatory drugs with cinnamyl alcohol are potent lipoxygenase inhibitors with enhanced anti-inflammatory activity. *Bioorg. Med. Chem. Lett.* **2015**, *25*, 5028–5031. [CrossRef] [PubMed]
26. Carnesecchi, S.; Schneider, Y.; Ceraline, J.; Duranton, B.; Gosse, F.; Seiler, N.; Raul, F. Geraniol, a component of plant essential oils, inhibits growth and polyamine biosynthesis in human colon cancer cells. *J. Pharmacol. Exp. Ther.* **2001**, *298*, 197–200. [PubMed]

© 2018 by the authors. Licensee MDPI, Basel, Switzerland. This article is an open access article distributed under the terms and conditions of the Creative Commons Attribution (CC BY) license (http://creativecommons.org/licenses/by/4.0/).

Review

Titanium Dioxide as a Catalyst in Biodiesel Production

Claudia Carlucci *, Leonardo Degennaro and Renzo Luisi

Flow Chemistry and Microreactor Technology FLAME-Lab, Department of Pharmacy–Drug Sciences, University of Bari "A. Moro" Via E. Orabona 4, 70125 Bari, Italy; leonardo.degennaro@uniba.it (L.D.); renzo.luisi@uniba.it (R.L.)
* Correspondence: claudia.carlucci@uniba.it; Tel.: +39-080-544-2251

Received: 30 November 2018; Accepted: 20 December 2018; Published: 11 January 2019

Abstract: The discovery of alternative fuels that can replace conventional fuels has become the goal of many scientific researches. Biodiesel is produced from vegetable oils through a transesterification reaction that converts triglycerides into fatty acid methyl esters (FAME), with the use of a low molecular weight alcohol, in different reaction conditions and with different types of catalysts. Titanium dioxide has shown a high potential as heterogeneous catalyst due to high surface area, strong metal support interaction, chemical stability, and acid–base property. This review focused on TiO_2 as heterogeneous catalyst and its potential applications in the continuous flow production of biodiesel. Furthermore, the use of micro reactors, able to make possible chemical transformations not feasible with traditional techniques, will enable a reduction of production costs and a greater environmental protection.

Keywords: titanium dioxide; heterogeneous catalyst; biodiesel; continuous flow

---

1. Introduction

Public attention to energy consumption and related emissions of pollutants is growing. The constant increase in the cost of raw materials derived from petroleum and the growing concerns of environmental impact have given considerable impetus to new products research from renewable raw materials and to technological proposal solutions that reduce energy consumption, use of hazardous substances and waste production, while promoting a model of sustainable development and social acceptance [1–3].

In recent years, titanium complex catalytic systems consisting of several catalysts or containing one catalyst with functional additives have found wide applications [4–6]. This application is very promising, since it appreciably widens the possibility of controlling the activity and selectivity of catalysts.

Titanium containing catalysts can be divided into organic, inorganic, mixed, and complex catalysts. Both organic and inorganic titanium compounds represent the main components of the complex catalysts for esterification and transesterification reactions [7].

Recently, titanium oxide (TiO_2) was introduced as an alternative material for heterogeneous catalysis due to the effect of its high surface area stabilizing the catalysts in its mesoporous structure [8].

Titania-based metal catalysts have attracted interest due to TiO_2 nanoparticles high activity for various reduction and oxidation reactions at low pressures and temperatures. Furthermore, TiO_2 was found to be a good metal oxide catalyst due to the strong metal support interaction, chemical stability, and acid–base property [9].

This review focuses on TiO_2 as an excellent material for heterogeneous catalysis, with potential applications in biodiesel production. Applications of titanium dioxide as heterogeneous catalyst for continuous flow processes have been considered.

2. Titania-Based Catalysts in Transesterification Reaction

Homogeneous basedcatalysts in the transesterification reaction have some disadvantages, among which are high energy consumption, expensive separation of the catalyst from the reaction mixture, and the purification of the raw material. Therefore, to reduce the cost of the purification process, heterogeneous solid catalysts such as metal oxides were recently used, as they can be easily separated from the reaction mixture and reused.

Titanium dioxide used as a heterogeneous catalyst shows a wide availability and economical synthesis modalities.

2.1. Sulfated TiO_2

A solid superacidic catalyst used in the petrochemical industry and petroleum refining process was sulfated doped TiO_2 [10,11]. This catalyst showed better performances compared to other sulfated metal oxides due to the acid strength of the TiO_2 particles which further enhanced with loading of SO_4^{2-} groups on the surface of TiO_2. The higher content of sulfate groups determined the formation of Brönsted acid sites which caused the super acidity of the catalyst [12]. Some studies reported that the enhancement of the acidic properties after the addition of sulfate ions to metal oxides, caused less deactivation of the catalyst [13,14].

2.1.1. SO_4^{2-}/TiO_2

Hassanpour et al. described sulfated doped TiO_2 as a solid super-acidic catalyst which is also used in the petrochemical industry and petroleum refining process and shows better performances compared to other sulfated metal oxides [15,16]. This is due to the acid strength of the TiO_2 particles which further enhanced with loading of SO_4^{2-} groups on the surface of TiO_2. The synthesized nano-catalyst $Ti(SO_4)O$ (Figure 1) is used for the production of biodiesel deriving from used cooking oil (UCO).

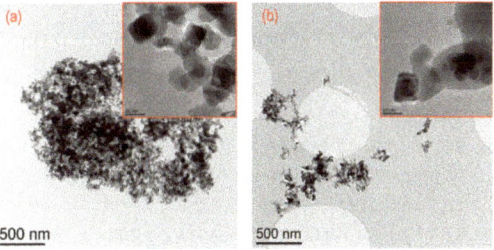

Figure 1. TEM images of (**a**) TiO_2 and (**b**) $Ti(SO_4)O$ (Copyright of Elsevier, see [15]).

The esterification of free fatty acids (FFAs) and transesterification of oils were conducted simultaneously using the titanium catalyst (1.5 wt.%), in methanol/UCO in 9:1 ratio, a temperature reaction of 75 °C, and a reaction time of 3 h, yielding 97.1% of fatty acid methyl esters. The authors investigated the catalytic activity and re-usability of the $Ti(SO_4)O$ for the esterification/transesterification of UCO. After eight cycles under optimized conditions the amount of SO_4^{2-} species in the solid acid nano-catalyst slowly decreased and this data resulted higher compared to other functionalized titania reported in the literature. The formation of polydentate sulfate species inside the structure of TiO_2 enhanced the stability of synthesized $Ti(SO_4)O$ nanocatalyst and also presented a higher tolerance to ≤6 wt.% percentage of free fatty acids in raw material for biodiesel production (Table 1).

Table 1. The effect of free fatty acid (FFA) in feedstock on the percentage of fatty acid methyl esters (FAME) yield.

Oleic acid to oil, wt.%	0.5	1	2	3	4	5	5.5	6	6.5	7
FAME yield%	97.1	97	97.1	97.01	96.14	95.69	93.42	91.37	75.39	64.5

Zhao and co-workers have recently studied the catalytic activity of sulfated titanium oxide (TiO_2-SO_4^{2-}). The authors reported that the high surface acidity of titanium dioxide increased the yield of butyl acetate to about 92.2% in esterification reaction, and the selectivity of the catalyst mostly depended on the degree of exposure of reactive crystal facets [17]. In this paper, a high-surface-area mesoporous sulfated nano-titania was prepared by a simple hydrothermal method without any template followed by surface sulfate modification (Figure 2). Acid sites with moderate- and superacidic strength formed in the sulfated titania catalyst. Also, the prepared sulfated sample possessed both Lewis and Brønsted acid sites. The catalytic activity of sulfated nano-titania with exposed (101) facets was evaluated using the esterification reaction between acetic acid and n-butanol. Compared with the exposed (001) facets, the exposed (101) facets showed better catalytic activity of sulfated TiO_2 in esterification. Additionally, the as-prepared sulfated sample could be efficiently recycled and regenerated by simple soaking in sulfuric acid followed by calcination.

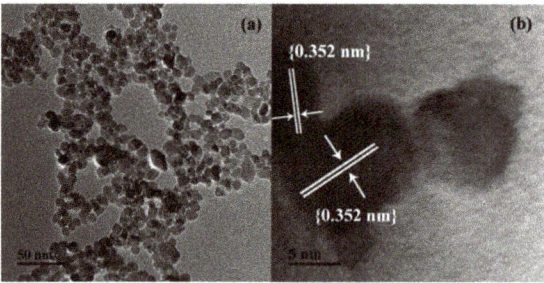

Figure 2. (a) TEM and (b) HRTEM images of TiO_2-SO_4^{2-} (Reproduced by permission of The Royal Society of Chemistry, see [17]).

Furthermore, Ropero-Vega et al. investigated the effect of TiO_2-SO_4^{2-} on the esterification of oleic acid with ethanol [18]. The maximum conversion of oleic acid was 82.2%, whilst a 100% selectivity of the catalyst on oleic acid to ester was reported at 80 °C after 3 h. Sulfated titania was prepared by using ammonium sulfate and sulfuric acid as sulfate precursors. Depending on the sulfation method, important effects on the acidity, textural properties as well as on activity were found. After ammonium sulfate was used, a large amount of S=O linked to the titania surface was observed and the acidity strength determined with Hammett indicators showed strong acidity in the sulfated samples. The presence of Lewis and Brønsted acid sites in the sulfated titania with sulfuric acid catalyst, were observed (Figure 3). The sulfated titania showed very high activity for the esterification of fatty acids with ethanol in a mixture of oleic acid (79%). Conversions up to 82.2% of the oleic acid and selectivity to ester of 100% were reached after 3 h of reaction at 80 °C.

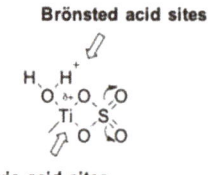

Figure 3. Schematic representation of the Brönsted and Lewis acid sites in the sulfated titania.

The results showed that sulfated titania is a promising solid acid catalyst to be used in the esterification of free fatty acids with 2-propanol (Table 2).

Table 2. Conversion of 2-propanol and oleic acid esterification on sulfated titania.

Catalyst	2-Propanol	Oleic Acid
[TiO_2-HNO_3]	0	3.1
[TiO_2/SO_4^{2-}-H_2SO_4-IS]	0.5	2.1
[TiO_2/SO_4^{2-}-$(NH_4)_2SO_4$-IS]	10.54	47.0
[TiO_2/SO_4^{2-}-$(NH_4)_2SO_4$-I]	46.06	82.2

Three sulfated titania-based solid superacid catalysts were prepared by sol-gel and impregnation method by Huang and coworkers [19]. Sulfated titania derived gel was dried at 353 K for 24 h and then calcined in air at 773 K for 3 h and milled into powders (this sample was labeled as ST). Another sulfated titania was prepared with HNO_3 instead of H_2SO_4 solution (this sample was labeled as HST). Sulfated titania-alumina was labeled as STA. The synthesis of biodiesel was performed from rap oil, at 353 K, after 6–12 h, under atmospheric pressure, with a 1:12 molar ratio of oil to methanol. The highest yield was obtained using HST catalyst probably due to its stronger surface acidity. The yields of HST and STA increased with prolonged reaction time, while the optimum reaction time of ST was 8 h (Figure 4).

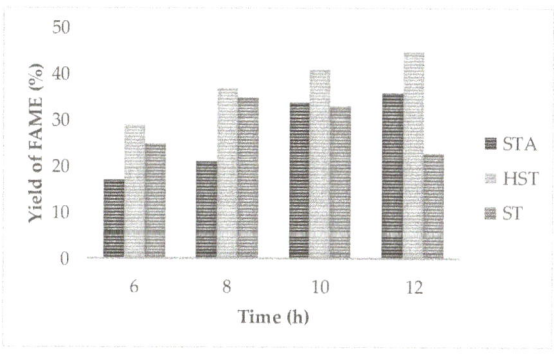

Figure 4. Influence of reaction time on the yield of FAME.

Superacid sulfated titania catalyst, TiO_2/SO_4 (TS-series), have been prepared by de Almeida et al. via the sol-gel technique, with different sulfate concentrations [20]. The relation of structure and catalytic activity of the prepared material have been evaluated. The catalyst that exhibited the highest catalytic activity in the methanolysis of soybean and castor oils at 120 °C, for 1 h (40% and 25%, respectively) was that which displayed the highest specific surface area, average pores diameter and pore volume, and highest percentage in sulfate groups TS-5 (Figure 5).

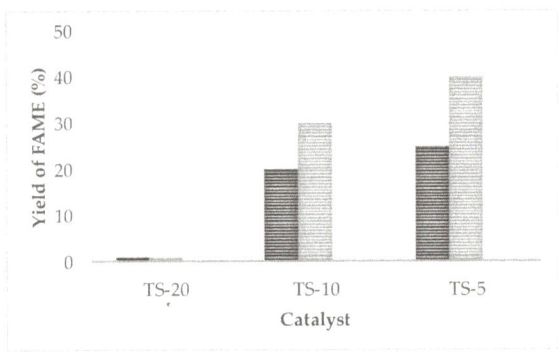

Figure 5. Percentage of FAMEs formed in the methanolysis reaction of soybean and castor oils.

Chen, et al. reported the transesterification reaction of cottonseed oil at 230 °C for 8 h, using a molar ratio of 12:1 between methanol and oil and an amount of catalyst of 2 wt.%, with a biodiesel conversion of 90% (Table 3) [21]. The solid acids as heterogeneous catalysts showed high activity for the transesterification and better adaptability compared to solid base catalysts in presence of a high acidity of the oil. The solid acid catalysts were prepared by mounting H_2SO_4 on $TiO_2 \cdot nH_2O$ and calcinated at 550 °C.

Table 3. Effect of temperature on the transesterification with catalyst TiO_2/SO_4^{2-} (%, w).

Temperature (°C)	Triglycerides	Diglycerides	Monoglycerides	Methyl Esters
200	4.7	4.8	4	86.5
210	3.2	3.7	3.1	90
220	2.3	2.5	2	93.2
230	1	1.8	1.2	96

2.1.2. SO_4^{2-}/TiO_2-ZrO_2

Oprescu et al. reported an alternative source for biodiesel production, using microalgae as source of oil and an amphiphilic SO_4^{2-}/TiO_2-ZrO_2 superacid catalyst and transesterification over KOH [22]. The extracted oil presented high free fatty acids (FFA) and required pre-treatment, if homogeneous catalysts were used due to saponification phenomenon and post-production processes. The biodiesel was obtained by transesterification over KOH and esterification of FFA with methanol using the amphiphilic SO_4^{2-}/TiO_2-ZrO_2 superacid catalyst. SO_4^{2-}/TiO_2-ZrO_2 was prepared with an alkylsilane to modify the surface of the catalyst. The attachment of alkylsilane on the surface of SO_4^{2-}/TiO_2-ZrO_2 support was confirmed by FT-IR and thermo gravimetric analysis. The authors evaluated the catalytic performance varying reaction parameters such as amount of catalyst, reaction time and algal oil/alcohol molar ratio (Table 4). To reduce algae oil acidity to less than 1% the acid esterification was carried out and, after transesterification with KOH, the yield of biodiesel was over 96%.

Table 4. Parameters optimization for esterification reaction over SO_4^{2-}/TiO_2-ZrO_2.

Time Reaction (h)	Catalyst Loading (wt.% Algae Oil)	Molar Ratio Algae Oil:Methanol	Acidity
5	6	1.3	4.98
5	6	1.6	2.23
5	6	1.9	1.57
5	6	1.12	1.60
5	2	1.9	5.40
5	4	1.9	3.34
5	6	1.9	1.57
5	8	1.9	1.61

Table 4. *Cont.*

Time Reaction (h)	Catalyst Loading (wt.% Algae Oil)	Molar Ratio Algae Oil:Methanol	Acidity
1	6	1.9	3.78
2	6	1.9	2.55
3	6	1.9	2.09
4	6	1.9	1.89
5	6	1.9	1.57
6	6	1.9	1.58

Boffito et al. described the preparation of different samples of sulfated mixed zirconia/titania, with traditional- and ultrasound (US)-assisted sol-gel synthesis, and the corresponding properties in the free fatty acids esterification [23]. The acidity and the surface area of sulfated zirconia was increased through the addition of TiO_2 and the same properties with the continuous or pulsed US were also tuned (Table 5). Furthermore, specific values of acidity and surface area were combined to demonstrate which kind of active sites were involved to obtain better catalytic performances in the free fatty acids esterification. SZ and SZT, referred to SO_4^{2-}/ZrO_2 and $SO_4^{2-}/80\%ZrO_2$-$20\%TiO_2$, were synthesized using traditional sol-gel method and both traditional and US assisted sol–gel techniques, respectively, while samples named USZT referred to US obtained sulfated $80\%ZrO_2$-$20\%TiO_2$ (Figure 6).

Table 5. List of all samples and of employed synthesis parameters (maximum power = 450 W).

Sample	Synthesis Time	Sonication Time	Acid Capacity (meq H$^+$/g)	Specific Surface Area (m^2 g^{-1})
SZ	123'0''	0''	0.30	107
SZT	123'0''	0''	0.79	152
SZT_773_6h	123'0''	0''	0.21	131
USZT_20_1_30	43'0''	43'0''	0.92	41.7
USZT_40_0.1_30	43'0''	4'18''	1.03	47.9
USZT_40_0.3_30	43'0''	12'54''	1.99	232
USZT_40_0.5_7.5	17'30''	8'45''	1.70	210
USZT_40_0.5_15	26'0''	13'0''	2.02	220
USZT_40_0.5_30	43'0''	21'30''	2.17	153
USZT_40_0.5_60	77'0''	38'30''	0.36	28.1
USZT_40_0.7_30	43'0''	30'6''	1.86	151
USZT_40_1_15	26'0''	26'0''	3.06	211
USZT_40_1_30	43'0''	43'0''	1.56	44.1

Figure 6. Conversions obtained after 6 h of reaction, 336 ± 2 K, slurry reactor, initial acidity: 7.5 wt.% (oleic acid), MeOH:oil = 16:100 wt, catalyst:oleic acid = 5:100 wt.

2.1.3. SO_4^{2-}/TiO_2-SiO_2

Wang et al. reported a study on the use of SO_4^{2-}/TiO_2-SiO_2 as a solid acid catalyst for the simultaneous esterification and transesterification of low cost feedstocks with high FFA [24]. The authors reported that with a mixed oil (50% refined cottonseed oil and 50% oleic acid), under 9:1 methanol to oil moral ratio, 6 h reaction time, 3% catalyst loading, and reaction temperature of 200 °C, a yield of 92% can be achieved. It was also reported that the SO_4^{2-}/TiO_2-SiO_2 catalyst can be re-used up to 4 times without reducing the catalytic activity (Figure 7).

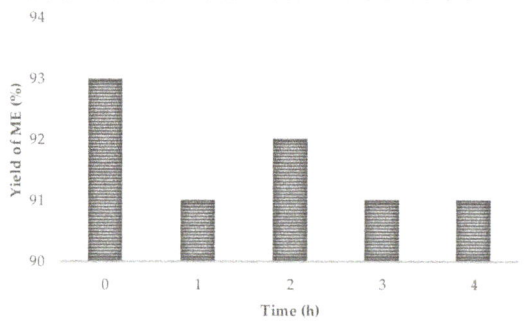

Figure 7. Stability of the solid acid catalyst.

Recently, an inexpensive precursor was used in the synthesis of SO_4^{2-}/TiO_2-SiO_2 catalyst by Shao and co-workers [25]. They reported 88% yield for biodiesel production under 20:1 methanol to UCO molar ratio, 10 wt.% catalyst and 3 h reaction time at 120 °C with constant stirring at 400 rpm. A sulfated titania–silica composite (S-TSC) was obtained through surface modification of mesoporous titania–silica composite synthesized using less expensive precursors; titanium oxychloride and sodium silicate as titania and silica sources respectively. A preformed titania sol facilitated the synthesis of a mesoporous composite, suitable for surface modification using sulfuric acid to improve its catalytic performance. FT-IR analysis showed the vibration band, not prominent, of the TiAOASi bond at 943 cm^{-1}, suggesting the incorporation of titania into silica to form a composite. This vibration band was substantially shifted to 952 cm^{-1} after the attachment of the sulfate group (Figure 8a). In the FT-IR spectrum of sulphated titania, calcined at 450 °C, new peaks were observed at 1043–1125 cm^{-1} attributable to the presence of the sulfate group (Figure 8b).

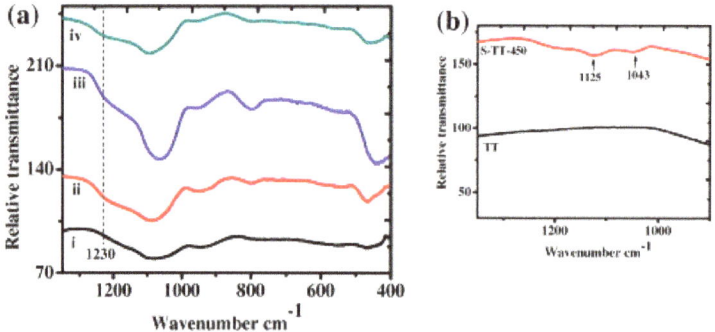

Figure 8. FT-IR spectra of the pure and sulfated titania–silica catalysts (**a**) titania–silica composite (TSC) (i), sulfated titania–silica composite (S-TSC) (ii), S-TSC-450 (iii), S-TSC-550 (iv). (**b**) The FT-IR spectra of pure and calcined sulfated titania (Copyright of Elsevier, see [25]).

The catalytic activity of a series of as-prepared TSC, S-TSC calcined samples and pure H_2SO_4 were evaluated for esterification of oleic acid and transesterification of waste oil with methanol to yield methyl esters (Table 6). It was observed that at these reaction conditions, S-TSC-450 and S-TSC-550 possessed high catalytic activity comparable to that of pure H_2SO_4 implying that surface modification of the titania–silica composite improved its acidic properties.

Table 6. Reaction of different catalyst in oleic acid esterification and waste oil transesterification.

Catalyst 10% (wt)	Conversion % OA	Conversion % WO
H_2SO_4	91.6	94.7
TSC-550	29.6	2.6
S-TSC-450	93.7	77
S-TSC-550	93.8	70.4
S-TSC-650	37.3	12.2
S-TSC-800	9.8	Not active
S-TT-450	93.4	88.1

Reaction conditions: 120 °C, 3 h, molar ratio MeOH/reagent 20/1.

Maniam et al. have recently used SO_4^{2-}/TiO_2-SiO_2 catalyst for the transesterification of decanter cake produced from waste palm oil into biodiesel. It was found that 120 °C reaction temperature, 1:15 oil to methanol ratio, 5 h transesterification time, and 13 wt.% catalyst loading, yielded a 91% of biodiesel [26]. Decanter cake (DC) was a solid waste produced after centrifugation of the crude palm oil. The pure palm oil was the supernatant while the decanter cake was the sediment. A high free fatty acids (FFA) content of DC-oil can be subjected to esterification, together with the transesterification of triglycerides.

2.1.4. $SO_4^{2-}/TiO_2/La^{3+}$

A solid acid catalyst $SO_4^{2-}/TiO_2/La^{3+}$ catalyzed both the esterification and transesterification of waste cooking oil with high content of free fatty acids (Figure 9) [27].

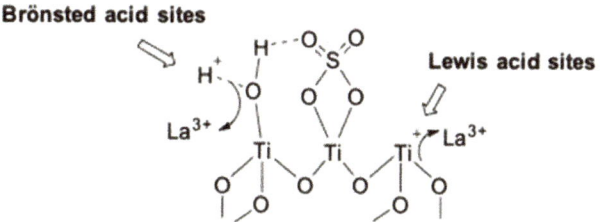

Figure 9. The framework structure of $SO_4^{2-}/TiO_2/La^{3+}$ catalyst.

Under the optimized conditions (catalyst amount 5 wt.% of oil, 10:1 molar ratio methanol to oil, temperature 110 °C and reaction time of 1 h) biodiesel was obtained with more than 90% of yield. The catalyst exhibited high activity after five cycles by activation and the content of fatty acid methyl esters was 96.16% (Table 7).

Table 7. FAMEs yield (%) with various catalyst reused times over catalysts.

Catalyst	RT1 [a]	RT2 [b]	RT3 [c]	RT4 [d]	RT5 [e]
SO_4^{2-}/TiO_2 (ST)	73.3	57.1	39.5	Trace	Trace
SO_4^{2-}/TiO_2-SiO_2 (STS)	80.1	78.6	75.0	70.8	61.6
$SO_4^{2-}/TiO_2/La^{3+}$ (STL)	92.3	92.1	91.7	91.1	90.2

[a] reused one time; [b] reused two times; [c] reused three times; [d] reused four times; [e] reused five times.

A new SO_4^{2-}/TiO_2-ZrO_2 solid superacid catalyst loaded with lanthanum was prepared by Li and coworkers [28]. They studied the catalytic performance for the synthesis of fatty acid methyl ester from fatty acid and methanol. The optimized conditions for the preparation of the catalyst were 0.1 wt.% amount of $La(NO_3)_3$, 0.5 mol^{-1} of the concentration of H_2SO_4 and 550 °C of calcination temperature. A conversion yield of 95% was reached after 5 h at 60 °C, with a catalyst amount of 5 wt.% and methanol amount of 1 mL/g fatty acid (FA). After five cycles the catalyst can be reused without any treatments and the conversion efficiency remained still at 90% (Table 8).

Table 8. Stability of the catalyst.

Catalyst	Reaction Cycles				
	1 (%)	2 (%)	3 (%)	4 (%)	5 (%)
SO_4^{2-}/TiO_2-ZrO_2/La^{3+}	97.8	95.9	95.8	95.1	93.6
SO_4^{2-}/TiO_2-ZrO_2	86.9	82.5	80.7	73.1	65.2

2.1.5. $SO_4^{2-}/TiO_2/Fe_2O_3$

Viswanathan and coworkers synthesized sulfated Fe_2O_3/TiO_2 (SFT) calcined over 300–900 °C [29]. The authors studied the transesterification of soybean oil with methanol varying sulfate contents over unsulfated and sulfated Fe_2O_3/TiO_2 catalysts and evaluating the acidity (Figure 10).

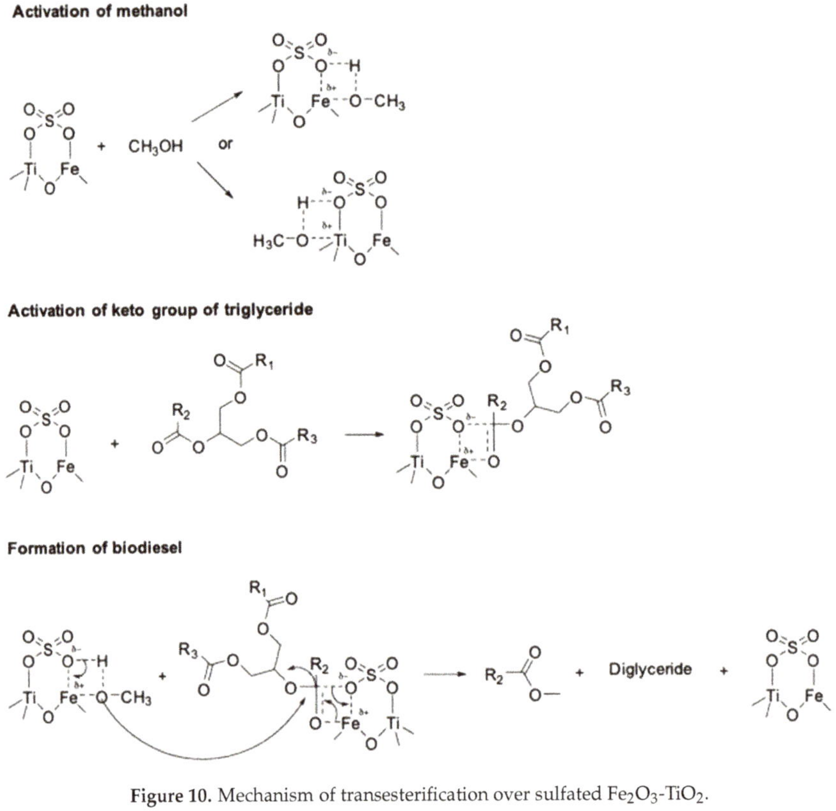

Figure 10. Mechanism of transesterification over sulfated Fe_2O_3-TiO_2.

The catalysts calcinated below 500 °C showed higher conversion of vegetable oil and significant yield of biodiesel probably due to the greater affinity of hydroxyl groups of methanol on Fe_2O_3/TiO_2. The removal of sulfate groups during calcination over 500 °C probably decreased the yield of biodiesel (Table 9).

Table 9. Products and yields of reactions with unsulfated and sulfated Fe_2O_3/TiO_2 catalyst.

Sample	Soybean Oil Conv. (%)	Monoglyceride Fatty Acids (%)	Diglyceride (%)	Triglyceride (%)	Biodiesel (%)
FT-500 [a]	23.6	4.20	16.6	76.4	2.80
SFT-300	100	7.54	1.06	Traces	91.4
SFT-500	98.3	5.11	1.01	1.68	92.2
SFT-700	76.5	15.5	18.3	20.5	45.7
SFT-900	65.7	11.3	24.2	34.3	30.2

Reaction conditions: sample 1.5 g; methanol to oil ratio 1:20; temperature 373 K; time 2 h. [a] Time 5 h.

2.2. TiO_2-Supported-ZnO_4

Afolabi and coworkers studied the catalytic properties of 10 wt.% of mixed metal oxide TiO_2-supported-ZnO catalyst. The conversion of waste cooking oil into biodiesel was investigated at 100, 150, and 200 °C, after 1 h, in the presence of methanol and hexane as co-solvent, with hexane to oil ratio of 1:1 [30]. Reaction time and temperature increased the biodiesel conversion from 82% to 92% while using hexane as co-solvent increased the rate of transesterification reaction producing higher biodiesel yields in shorter time.

Piraman and coworkers used mixed oxides of TiO_2-ZnO and ZnO catalysts as active and stable catalysts for the biodiesel production [31]. 200 mg of TiO_2-ZnO catalyst loading exhibited good catalytic activities, a 98% conversion of fatty acid methyl esters was achieved with 6:1 methanol to oil molar ratio, in 5 h, at 60 °C. The catalytic performance of TiO_2-ZnO mixed oxide was better compared to ZnO catalyst, and this catalyst can be used for the large-scale biodiesel production (Figure 11).

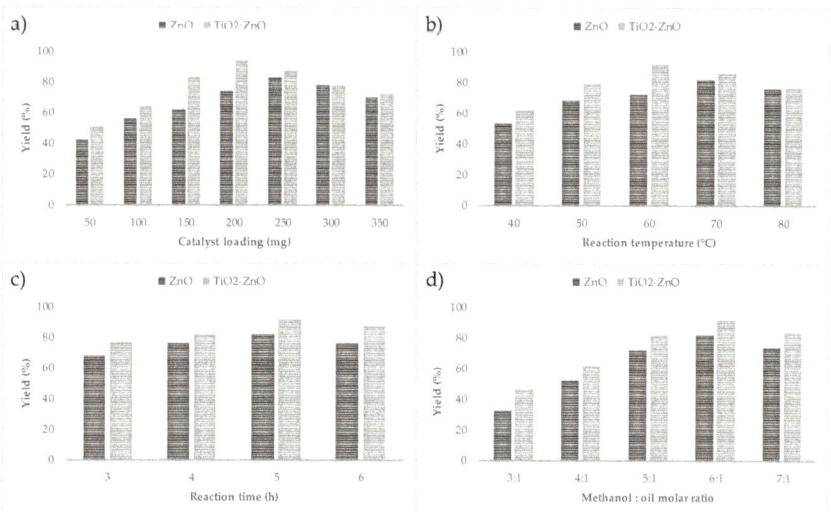

Figure 11. Effect of experimental parameters on FAME conversion: (**a**) catalyst loading, (**b**) temperature, (**c**) time and (**d**) methanol/oil molar ratio.

2.3. TiO$_2$-MgO

Kalala and coworkers reported the preparation of titania supported MgO catalyst samples (10 and 20 wt.% MgO loading) tested as catalyst for the conversion of waste vegetable oil to biodiesel in presence of methanol, with an alcohol to oil molar ratio of 18:1 [32]. The effects of reaction temperature and reaction time increased the oil conversion while the effect of MgO loading on the waste oil conversion depended on the operating temperature. After 1 h, at 60, 150, 175, and 200 °C the resulting conversion yields were 42, 55, 86, and 89% respectively, using a 20 wt.% of MgO loading.

In another work, nano-MgO was deposited on titania using deposition-precipitation method and its activity was tested on the transesterification reaction of soybean oil to biodiesel [33]. The catalyst activity was improved increasing the reaction temperature from 150 and 225 °C while increasing the reaction time over 1 h significant conversion was not observed. The authors investigated the stability of MgO on TiO$_2$ and they observed a MgO loss during the reaction between 0.5 and 2.3 percent, without correlation between the reaction temperature.

Wen et al. used mixed oxides of MgO-TiO$_2$ (MT) produced by the sol-gel method to convert waste cooking oil into biodiesel [34]. The best catalyst was MT-1-923 comprising a Mg/Ti molar ratio of 1 and calcined at 650 °C. The authors investigated the main reaction parameters such as methanol/oil molar ratio, catalyst amount and temperature. The best yield of FAME 92.3% was obtained at a molar ratio of methanol to oil of 50:1; catalyst amount of 10 wt.%; reaction time of 6 h and reaction temperature of 160 °C. They observed that the catalytic activity of MT-1-923 decreased slowly in the recycle process. To improve catalytic activity, MT-1-923 was regenerated by a two-step washing method (the catalyst was washed with methanol four times and subsequently with n-hexane once before being dried at 120 °C). The FAME yield slightly increased to 93.8% compared with 92.8% for the fresh catalyst due to an increase in the specific surface area and average pore diameter. Titanium improved the stability of the catalyst because of the defects induced by the substitution of Ti ions for Mg ions in the magnesia lattice. The best catalyst was determined to be MT-1-923, which is comprised of an Mg/Ti molar ratio of 1 and calcined at 923 K, based on an assessment of the activity and stability of the catalyst. The main reaction parameters, including methanol/oil molar ratio, catalyst amount, and temperature, were investigated (Table 10).

Table 10. Effects of reaction parameters on the performance of the MT-1-923.

Methanol/Oil (Molar Ratio)	Catalyst Amount (wt.%)	Temperature (K)	Biodiesel Yield [a] (%)
20	5	423	52
30	5	423	79.9
40	5	423	83.5
50	5	423	85.6
60	5	423	85.3
50	6	423	86.9
50	8	423	86.9
50	10	423	91.2
50	12	423	91.2
50	15	423	89.3
50	10	403	22.3
50	10	413	67.6
50	10	433	92.3
50	10	443	91.6

[a] Reaction conditions: reaction time 6 h, stirring speed 1500 rpm.

2.4. CaTiO$_3$

Kawashima and coworkers investigated the transesterification of rapeseed oil using heterogeneous base catalysts [35]. They prepared different kinds of metal oxides containing calcium, barium or magnesium and tested the catalytic activity at 60 °C, a reaction time of 10 h and with a 6:1 molar ratio of methanol to oil. The calcium-containing catalysts CaTiO$_3$, CaMnO$_3$, Ca$_2$Fe$_2$O$_5$, CaZrO$_3$, and CaCeO$_3$ showed high activities and yields of biodiesel conversion (Table 11).

Table 11. Surface area and catalytic activities of metal oxides.

Sample	Surface Area (m²/g)	Methyl Ester Yield (%)
CaTiO$_3$	4.9	79
CaMnO$_3$	1.5	92
Ca$_2$Fe$_2$O$_5$	0.7	92
CaZrO$_3$	1.8	88
CaCeO$_3$	2.9	89
BaZrO$_3$	3.3	0.4
BaCeO$_3$	2.8	-
MgZrO$_3$	7.4	0.5
MgCeO$_3$	7.7	0.4

2.5. K-Loading/TiO$_2$

Guerrero and coworkers studied the transesterification reaction of canola oil on titania-supported catalysts with varying loadings of potassium [36]. In a previous work they investigated 20% K-loading catalyst under air conditions and without any treatment before reaction, which achieved the total conversion to methyl esters (Figure 12).

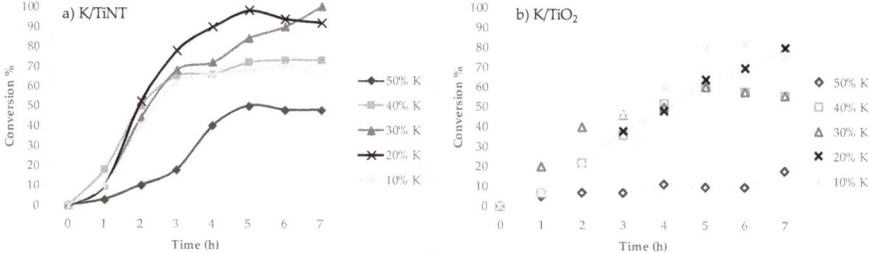

Figure 12. Conversion of canola oil to methyl esters with (**a**) K/TiNT and (**b**) K/TiO2 catalysts.

Afterwards they studied the transesterification reaction of canola oil for the biodiesel production using a hydrotreated TiO$_2$ supported potassium catalyst, K/TiHT [37]. The calcination at different temperatures led to the transformation of the supported potassium catalyst into a titanate form of oxide and this increased the activity of the catalyst. The recovery of the catalyst was then used in successive reactions leading to stable conversions and a maximum conversion was achieved with the optimum reaction conditions using a catalyst loading of 6% (w/w), a methanol to oil ratio of 54:1, and a temperature of reaction of 55 °C, with a catalyst calcined at 700 °C.

In a work by Klimova et al., sodium titanate nanotubes (TNT) doped with potassium were synthesized by the Kasuga method and tested as catalysts for biodiesel production [38]. To increase the basicity of the catalyst, potassium was added to the nanotubes and the efficiency in the transesterification of soybean oil with methanol was improved. To increase potassium loadings in the nanotubes the NaOH/KOH molar ratio was turned from 9:1 to 7:3. Sodium trititanate nanotubes containing 1.5 wt.% of potassium were obtained using a NaOH/KOH molar ratio of 9:1, with a 10 M alkali solution. Titanate nanotubes with larger potassium loadings (3.2 and 3.3 wt.%) were obtained increasing the proportion of KOH to 20 and 30 mol.% in the NaOH/KOH solutions. Potassium-containing nanotubes showed higher catalytic activity in the transesterification reaction compared to the pure sodium used as a reference. The best results were obtained at 80 °C, after 1 h with the samples containing 3.2–3.3 wt.% of potassium obtaining a biodiesel conversion yield of 94–96% (Table 12).

Table 12. Catalytic activity of the NaK(X)TNT (X refers to the percentage of the potassium loaded) samples.

Sample	NaTNT	NaK(10)TNT	NaK(20)TNT	NaK(30)TNT
Conversion to biodiesel (%)	58.4	74.3	96.2	94.3
Kinematic viscosity (mm^2/s)	7.8	6.0	4.3	4.5

Table 13 summarized the data previously reported.

Table 13. Use of titanium dioxide as catalyst in batch production of biodiesel.

Oil Source	Catalyst	Reactor	Conditions	Yield %	Ref.
Waste	[Ti(SO$_4$)O]	glass batch	75 °C, 3 h, methanol/oil 9:1 catalyst 1.5 wt.%	97.1	[15]
Acetic acid, n-butanol	TiO$_2$-SO$_4^{2-}$	flask	120 °C, 150 min methanol/oil 1.2 catalyst 1.8 g	92.2	[17]
Oleic	[TiO$_2$/SO$_4^{2-}$-(NH$_4$)$_2$SO$_4$] [TiO$_2$/SO$_4^{2-}$-H$_2$SO$_4$] [TiO$_2$/SO$_4^{2-}$-(NH$_4$)$_2$SO$_4$-IS]	reflux condenser	80 °C, 3 h methanol/oil 10:1 catalyst 2 wt.%	82.2	[18]
Rapeseed	TiO$_2$-SO$_4^{2-}$	flask	353 K, 6–12 h, methanol/oil 12:1 catalyst 1.2 g	51	[19]
Soybean Castor	TiO$_2$/SO$_4^{2-}$	stainless steel batch	120 °C, 1 h methanol/oil/catalyst 120:20:1	40 25	[20]
Cottonseed	TiO$_2$-SO$_4^{2-}$	autoclave	230 °C, 8 h methanol/oil 12:1 catalyst 2 wt.%	90	[21]
Microalgae	SO$_4^{2-}$/TiO$_2$-ZrO$_2$	flask	5 h, methanol/oil 9:1 catalyst 6 wt.%	96	[22]
Rapeseed	SO$_4^{2-}$/80%ZrO$_2$-20%TiO$_2$	oil bath	6 h at 336 ± 2 K, methanol/oil 4.5:1 catalyst 5 wt.%	42.4	[23]
Waste	SO$_4^{2-}$/TiO$_2$-SiO$_2$	autoclave	200 °C, methanol/oil 9:1 catalyst 3 wt.%	92	[24]
Waste Oleic acid	TiO$_2$-SiO$_2$ SO$_4^{2-}$/TiO$_2$-SiO$_2$	autoclave	120 °C, 3 h methanol/oil 20:1 catalyst 10 wt.%	94.7 93.8	[25]
Waste	SO$_4^{2-}$/TiO$_2$-SiO$_2$	reaction flask	120 °C, 5 h methanol/oil 15:1 catalyst 13 wt.%	91	[26]
Waste	SO$_4^{2-}$/TiO$_2$/La^{3+}	autoclave	110 °C, 1 h methanol/oil 10:1 catalyst 5 wt.%	96.16	[27]
Rapeseed	SO$_4^{2-}$/ZrO$_2$-TiO$_2$/La^{3+}	flask	60 °C, 5 h 5 wt.%, methanol 1 mL/g fatty acid (FA)	95	[28]
Soybean	sulfated Fe$_2$O$_3$/TiO$_2$	autoclave	100 °C, 2 h, methanol/oil 20:1 catalyst 15 wt.%	92.2	[29]
Waste	TiO$_2$-ZnO	pressurized reactor	200 °C, 1 h, methanol/oil 18:1 catalyst 10 wt.%	82.1	[30]
Palm	TiO$_2$-ZnO	flask	60 °C, 5 h methanol/oil 6:1 catalyst 200 mg	98	[31]
Waste	TiO$_2$-MgO	stainless steel batch	225 °C, 1 h methanol/oil 18:1 catalyst 20 wt.%	100	[32]
Soybean	Nano-MgO TiO$_2$	stainless steel batch	225 °C, 1 h methanol/oil 18:1 catalyst 5 wt.%	84	[33]

Table 13. *Cont.*

Oil Source	Catalyst	Reactor	Conditions	Yield %	Ref.
Waste	TiO$_2$-MgO	stainless steel batch	423 K, 6 h, methanol/oil 30:1 catalyst 5 wt.%	92.3	[34]
Rapeseed	CaTiO$_3$	flask	60 °C, 10 h methanol/oil 6:1	90	[35]
Canola	K/TiNT K/TiO$_2$	glass batch	70 °C, 5 h methanol/oil 36:1 catalyst 20 wt.%	100	[36]
Canola	K/TiHT	glass batch	55 °C, 3 h methanol/oil 54:1 catalyst 20 wt.%	>90	[37]
Soybean	NaK(20)TNT	stainless steel batch	80 °C, 1 h NaOH/KOH 7:3 catalyst 3.2 wt.%	96.2	[38]

3. Titania-Based Catalysts in Continuous Flow Microreactors

Flow chemistry is currently widely applied in the preparation of organic compounds, drugs, natural products and materials in a sustainable manner. Microreactors and streaming technologies have played an important role in both academic and industrial research in recent years, offering a viable alternative to batch processes [39–42]. The use of continuous processes, within "micro or meso-reactors", allowed access to a wider profile of reaction conditions not accessible through the use of traditional systems. The microfluidic systems allowed an optimization of the reaction parameters, such as mixing, flow rate, and residence time. Furthermore, pressure and temperature can be easily controlled, in parallel with other conditions such as solvent, stoichiometry and work-up operations.

Most efforts in this area focused on the selection of effective catalysts for biodiesel conversion via transesterification. However, to scale up the biodiesel production, many researchers utilized continuous-flow regime to continuously convert lipids to biodiesel with preferable process design to solve the problems encountered during continuous operation [43,44].

The advantages and limitations of using catalyzed transesterification in conventional continuous-flow reactors could minimize mass transfer resistance and improve biodiesel conversion. Conventionally, due to the presence of multiple phases during the catalytic reaction, the mass transfer between reactants and catalysts, as well as the type of catalyst used are the two major factors that should be considered during the design of the reactor applied for the targeted conversion [45].

Joshi and coworkers outlined the catalytic thermolysis of Jatropha oil using a model fixed bed reactor (Figure 13), in a range of temperature between 340 and 420 °C and at liquid hourly space velocity (LHSV) of 1.12, 1.87, and 2.25 h^{-1} (Table 14) [46]. They synthesized Amorphous Alumino-Silicate heterogeneous catalysts, separately loaded with transition metal oxide such as titania (SAT). The distillation over the temperature range of 60–200 °C of the crude liquid mixture produced four fractions. The boiling point, specific gravity, viscosity and calorific value of the first fraction resembled the properties of petrol while the second and the third fractions resembled diesel.

Figure 13. Schematic diagram of lab scale fixed bed reactor.

Table 14. Experimental results of Jatropha oil cracking over titania (SAT).

Temperature (°C)	LHSV (h^{-1})			Liquid Crude (vol.%)			Gaseous Product (per L of Feed)			Distilled Biofuel (vol.%)			Condensed Water (vol.%)		
340	1.12	1.87	2.25	77	80	84	85	88	76	57	55	58	1.5	0.9	1.3
380	1.12	1.87	2.25	79	85	87	90	78	95	55	59	51	2.1	1.6	1
420	1.12	1.87	2.25	75	77	80	102	98	105	52	54	50	1.1	0.7	0.9

McNeff and coworkers developed a novel continuous fixed bed reactor process for the biodiesel production using a metal oxide-based catalyst [47]. Porous zirconia, titania and alumina micro-particulate heterogeneous catalysts were used in the esterification and transesterification reactions under continuous conditions, high pressure (2500 psi) and elevated temperature (300–450 °C). The authors described a simultaneous continuous transesterification of triglycerides and esterification of free fatty acids, with residence times of 5.4 s (Table 15). Biodiesel was produced from soybean oil, acidulated soapstock, tall oil, algae oil, and corn oil with different alcohols and the process can be easily scaled up for more than 115 h without loss of conversion efficiency.

Table 15. Biodiesel production condition for base modified titania catalysts.

Catalyst	BMT	BMT	BMT	UMT	None
Reactor volume (mL)	23.55	23.55	23.55	23.55	2.49
Preheater T (°C)	363	370	340	350	455
Column inlet T (°C)	348	359	343	353	445
Column outlet T (°C)	339	325	344	355	462
Initial T (°C)	209	184	247	247	203
Final T (°C)	94	63	76	76	57
Initial P (psi)	3.65	3.1	2.5	2.5	3.05
Final P (psi)	2.7	2.7	2.3	2.3	2.7
Molar ratio (alcohol/oil)	32.7	32.7	32.7	50.0	32.7
Total flow rate (mL/min)	15.904	15.904	15.904	15.904	17.808
Residence time (s)	56.9	56.9	56.9	56.9	5.4

BMT: Base Modified Titania, UMT: Unmodified Titania.

The biodiesel plant based on the Mcgyan process is reported in Figure 14. The use of two high pressure HPLC pumps was shown. The oil feedstock was filtered under high pressure before entering the heat exchanger and combining with methanol. Both the alcohol and lipid feedstock were pumped into a stainless steel heat exchanger. Afterwards the reactant streams were combined using a "T" and preheated before entering the thermostated fixed bed catalytic reactor. The temperature was controlled and the backpressure of the system was maintained through the use of a backpressure regulator.

In another work of the same authors a highly efficient continuous catalytic process to produce biodiesel from Dunaliella tertiolecta, Nannochloropsis oculata, wild freshwater microalgae, and macroalgae lipids was developed [48]. Porous titania microspheres and supercritical methanol were used as heterogeneous catalyst in a fixed bed reactor to catalyze the simultaneous transesterification and esterification of triacylglycerides and free fatty acids, into biodiesel (Table 16). The authors used a feedstock solution of algae, hexane-methanol (97:3 w/w) as carrier solvent. The solution was pumped with high pressure HPLC through an empty stainless steel reactor and the reactant stream passed through a heat exchanger. The reactants were pumped across the reactor with a 30 s residence time, at 340 °C, with 2250 psi front pressure and the backpressure of the system was maintained through the use of a backpressure regulator.

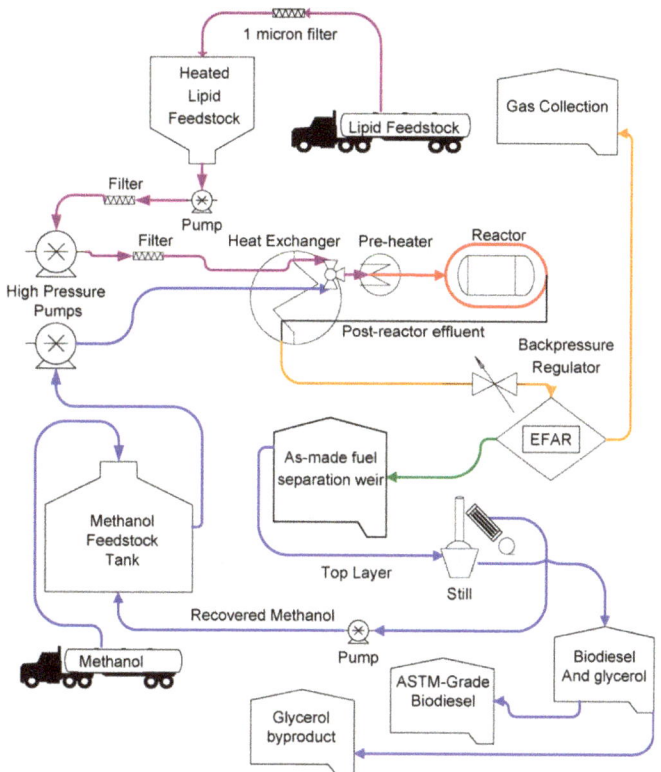

Figure 14. Process flow diagram of a biodiesel plant based on the Mcgyan process (Copyright of Elsevier, see [47]). EFAR: Easy Fatty Acid Removal.

Table 16. A comparison of the supercritical fixed–bed continuous flow process (the Mcgyan process) to the conventional homogeneously base catalyzed batch system.

	Mcgyan Process	Homogeneous Process
Consume of the catalyst	No	Yes
Large amounts of H_2O	No	Yes
Waste products	No	Yes
Soap byproducts	No	Yes
Glycerol as byproduct	No	Yes
Large footprint	No	Yes
Sensitive to H_2O	No	Yes
Sensitive to FFA	No	Yes
Large quantities of strong acid/base	No	Yes
Conversion rate	Sec	h/d
Variety of feedstocks	Yes	No
Continuous process	Yes	No

Aroua and coworkers described the production of biodiesel in a TiO_2/Al_2O_3 membrane reactor (Figure 15) [49]. The effects of reaction temperature, catalyst concentration and cross flow circulation velocity were investigated. Biodiesel was obtained through alkali transesterification of palm oil and separated in the ceramic membrane reactor at 70 °C, with 1.12 wt.% catalyst concentration and 0.211 cm s^{-1} cross flow circulation velocity. Palm oil and methanol were pumped into the system using two separate ways, with a methanol to oil ratio of 1:1 and various catalyst amounts were used in the packed membrane reactor. Methanol was charged and heated continuously into the reactor using a circulating pump afterwards the reactor was filled with palm oil. Pressure inside the membrane was monitored and after 60 min the circulating pump and heat exchanger were switched off and the products were collected into a separating funnel to separate biodiesel from glycerol. The ceramic membrane has shown an excellent chemical and physical stability even after one year of operation and contact with methanol and solid alkali catalyst.

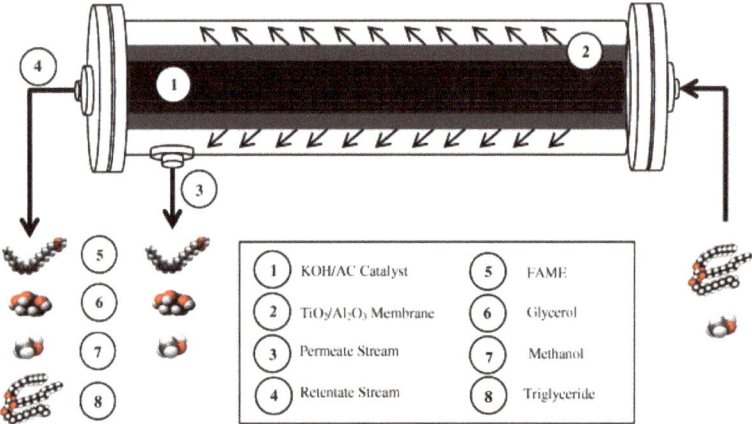

Figure 15. Combination of heterogeneous base transesterification and triglyceride separation in the packed bed membrane reactor (Copyright of Elsevier, see [49]).

Wang and coworkers proposed also a continuous process for biodiesel production from cheap raw feedstocks with high FFAs by solid acid catalysis (Figure 16) [24]. The production process was carried out pretreating the raw feedstocks by filtration and dehydration to remove impurities and water. In a series of three reaction boilers, part of the methanol reacted with oils as a reactant, and excess methanol removed water from the system as a solvent, which increased the esterification conversion substantially and effectively decreased the acid value. Finally, excess methanol was purified in a methanol distillation tower for recycling, while the oil phase was refined at a biodiesel vacuum distillation tower to give the biodiesel product. The proposed continuous process produced a 10,000-tonnes/year industrial biodiesel. The use of cheap feedstocks with high FFAs such as waste cooking oils, soapstocks, and non-edible oils, instead of refined vegetable oil, decreased the cost greatly. The solid catalyst SO_4^{2-}/TiO_2-SiO_2 had high catalytic activity, easy separation, and catalyzed biodiesel production by simultaneous esterification and transesterification.

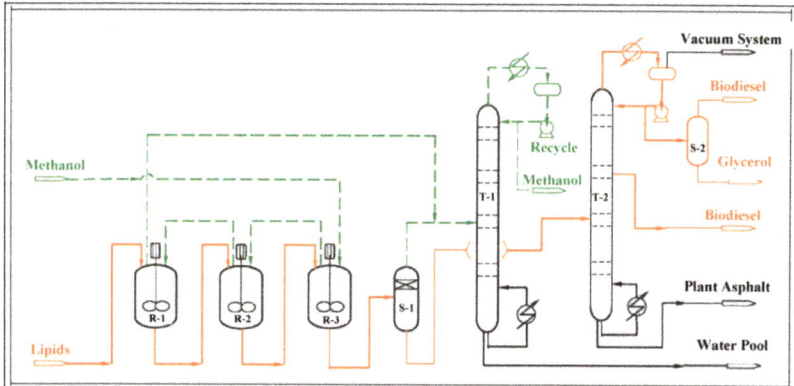

Figure 16. Process flow diagram of biodiesel production by solid acid catalysis (Copyright of Elsevier, see [24]).

The most abundantly produced waste of the biodiesel industrial production was glycerol and several studies of the continuous flow transformation of glycerol were reported in literature (Figure 17) [50].

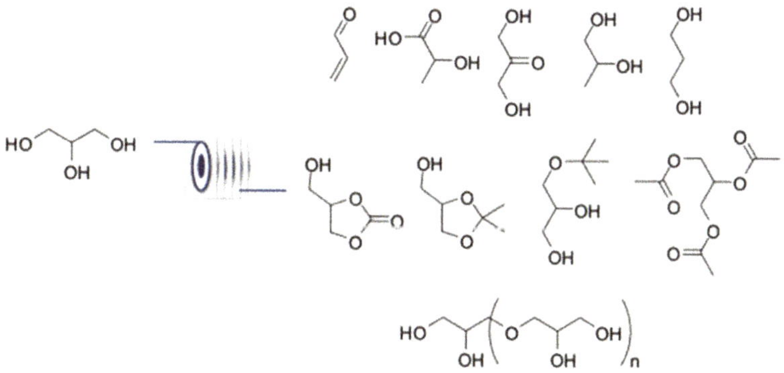

Figure 17. Roadmap of selected glycerol valorization reactions in continuous flow.

Chary and co-workers studied the hydrogenation of glycerol to 1- and 2-biopropanols under ambient condition and continuous process over a platinum catalyst supported on titanium phosphates (TiP) as supports (Figure 18) [51]. This heterogeneous catalyst shown excellent results as a steady 97% selectivity toward propanol (87:10 1-/2-propanol). Quantitative glycerol was recovered after 15 h of operation. The influence of reaction temperature on glycerol hydrogenolysis over the 2Pt/TiP catalyst was shown in Table 17. At 220 °C reaction temperature the maximum glycerol conversion and selectivity was obtained.

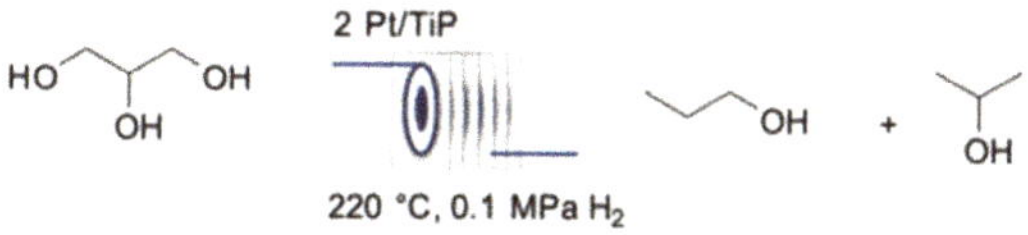

Figure 18. Hydrogenolysis of glycerol to produce 1- and 2-propanols.

Table 17. Effect of reaction temperature on glycerol hydrogenolysis over 2Pt/TiP catalyst.

Temperature (°C)	Conversion (%)	Total PO	1-PO [a]	2-PO [b]	1,2-PD	1,3-PD	Acrolein	Others [c]
180	70	74	59	15	9.0	3.2	9.3	4.5
200	82	80	67	13	5.4	2.2	8.7	3.7
220	100	97	87	10	–	–	02	01
240	100	88	80	08	–	–	4.2	7.8
260	100	84	76	08	–	–	01	15

[a] 1-propanol, [b] 2-propanol, [c] ethanol, ethylene glycol, hydroxyl acetone, methanol and acetone.

Paul and co-workers reported the dehydration of glycerol into acrolein using an alternative process using a flow sequence operating at high temperatures (280–400 °C) and atmospheric pressure (Table 18) [52]. The first step of dehydration of glycerol to acrolein was performed over WO_3/TiO_2 catalyst and the injection of molecular oxygen avoided the catalyst deactivation. The optimized conditions were a reaction temperature of 280 °C, a catalyst amount of 5.0 g; 11.33 mL/min O_2; contact time: 0.36 s; gas phase composition at 280 °C: 92.74% H_2O, 2.72% O_2, 4.54% glycerol.

Table 18. Catalytic performance of WO_3/TiO_2.

Time (h)	Glycerol (%)	Acrolein (%)	Hydroxyacetone (%)	Acetaldehyde (%)	Propionaldehyde (%)
1	97	40	1.1	1.7	7.2
2	98	61	1.1	1.9	1.1
3	97	82	1.9	1.3	0.5
4	93	84	1.6	1.5	0.6
5	95	81	2.2	1.2	0.5
Average	96	70	1.6	1.5	2.0

The indirect ammoxidation of glycerol was schematized in Figure 19. The reactions were performed under atmospheric pressure in continuous plug flow fixed-bed reactors, connected to a stainless steel evaporator used to vaporize the liquid reactants before contacting the catalytic bed, filled with granulated catalysts. In the ammoxidation of acrolein and glycerol, the products were condensed in an acidic solution to neutralize unreacted ammonia and to suppress polymerization of acrolein. The mass-flow controllers checked the flow rate of the gaseous reactants.

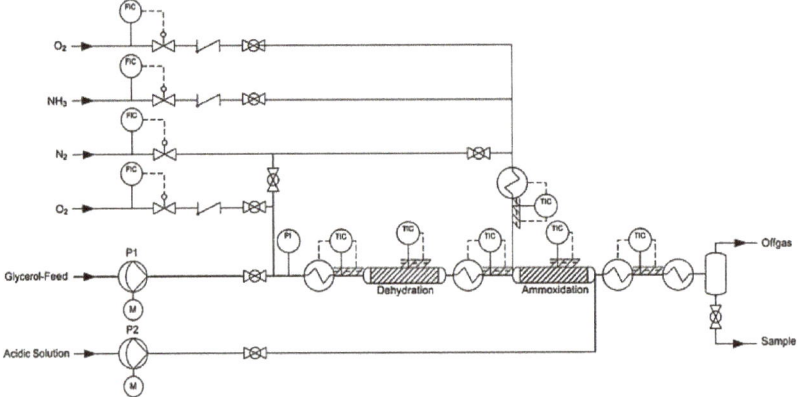

Figure 19. Reactor setup for the tandem process (Copyright of Elsevier, see [52]).

The results previously described are reported in Table 19.

Table 19. Use of TiO_2 as catalyst in continuous flow production of biodiesel and glycerol recovery.

Oil Sources	Catalyst	Reactor	Condition	Ref
Jatropha	Alumino-Silicate TiO_2	fixed bed	340 and 420 °C LHSV (1.12, 1.87, 2.25 h^{-1})	[46]
Soybean, tall, algae, acid soapstock, corn	TiO_2	fixed bed	2500 psi, 300–450 °C residence time 5.4 s	[47]
Dunaliella tertiolecta, Nannochloropsis oculata	TiO_2	fixed bed	2500 psi, 340 °C residence time 30 s	[48]
Palm	TiO_2/Al_2O_3	packed membrane	70 °C, 1.12 wt.% 0.211 cm s^{-1}	[49]
Source	Catalyst	Reactor	Condition	Ref
Glycerol	WO_3/TiO_2	plug flow fixed-bed	280 °C, WO_3/TiO_2 15 wt.% glycerol flow rate 23 g/h O_2 flow 11.33 mL/min	[52]

4. Titania-Based Additives to Biodiesel

The introduction of titanium dioxide nanoparticles as additive in biodiesel production has been less explored compared to the use of TiO_2 as catalyst. Nevertheless, the performances of biodiesel additivated with nano titania resulted increased and variations in properties such as kinematic viscosity, heating value, cetane number, specific fuel consumption, break thermal efficiency, indirect injection, smoke emissions, flash and fire point could be observed.

Jeryrajkumar investigated the effect of titanium dioxide nanoadditives on the performance and emission characteristics of Calophyllum innophyllum biodiesel (B100) in a single cylinder, four strokes, water cooled, compression injection diesel engine [53]. The nanoparticles were prepared by hydrothermal process with a size range of 100 nm. These additives reduced the fuel consumption, particulate matter (PM), carbon monoxide (CO) and unburned hydrocarbons (UHC) emissions while increased the nitrogen oxides (NOx) emission, the engine combustion, the thermal efficiency and performance characteristics. The performance and emission characteristics of a single cylinder, water cooled, compression ignition diesel engine with calophyllum inophyllum methyl ester using B100TiO_2 additives were investigated and compared with diesel. The nano additives blended biodiesel increased the brake thermal efficiency while the specific fuel consumption decreased. Titanium dioxide blended biodiesel increased gradually NOx emission at all loads, increased CO emission of 25% at the full load condition and reduced hydrocarbon (HC) emission of 70% at 75% of loading as compared with pure biodiesel.

Venu and coworkers investigated the biodiesel-ethanol (BE) blends in a compression ignition engine using a blend of BE with 25 ppm TiO_2 nanoparticle additives (denoted as BE-Ti). The addition of nanoparticles created large contact surface area with the base fuel, increased the combustion with minimal emissions and the oxidation rate while reduced the light-off temperature. The addition of titanium nanoparticles increased NOx, HC and smoke emissions while decreased Break Specific Fuel Consumption (BSFC) and CO emissions. Diethyl ether addition to BE blends increased improved the heat release rate, HC, CO emissions while decreased BSFC, NOx and smoke [54].

Binu and coworkers used titanium dioxide nanoparticles as fuel additive in a C.I. Engine. The nanoparticles were dispersed in B20 methyl ester of pongamia pinnata oil using a probe sonicator. The addition of titanium dioxide nanoparticles enhanced performance and emission characteristics of the fuel samples tested in C.I. Engine test. The use of TiO_2 nanoparticles in B20 blend increased brake thermal efficiency of the engine of 2% and reduced the BSFC value of 6% and smoke emission of 16% compared to plain B20 [55].

Prabhu and coworkers investigated the effect of titanium oxide (TiO_2) nanoparticle as additive for diesel-biodiesel blends on the performance and emission characteristics of a single cylinder diesel engine at different load conditions. Titanium oxide nanoparticles (250–500 ppm) were blended with 20% biodiesel–diesel (B20). Carbon monoxide (CO), hydrocarbon (HC) and smoke emissions were decreased while the brake thermal efficiency and the NO emissions were increased marginally for 250 ppm nano particle added with B20 blends when compared with B20 and 500 ppm added with B20 fuel at full load conditions [56].

Fangsuwannarak conducted a comparative study of palm biodiesel properties and engine performance efficiency with the addition of TiO_2 nanoparticles. The various palm oil fractions of 2, 10, 20, 30, 40, 50 and 100% in the rest of ordinary diesel fuel were denoted as B2, B10, B20, B30, B40, B50, and B100, respectively. The addition of nano-TiO_2 additive of 0.1 and 0.2% by volume was evaluated and it was found that the quality of the modified fuel with 0.1% TiO_2 increased cetane number, lower-heating value, and flash point while reduced kinematic viscosity. The performance of an indirect injection (IDI) engine and the control of carbon monoxide (CO), carbon dioxide (CO_2), and oxides of nitrogen (NOx) emissions were enhanced. The nano-TiO_2 additive of 0.1% by volume had the most effective performance of the tested engine for biodiesel blended fuel between B20 and B100 fuels. The additive 0.1% TiO_2 biodiesel fuel revealed the higher level of brake power, wheel power, and engine torque. Meanwhile, the level of specific fuel consumption significantly decreased. The effect on CO, CO_2, and NOx was also investigated in this study and it was demonstrated that 0.1% TiO_2 reduced the exhaust emissions and it is the most effective in B30 fuel [57].

Umashankar presented the effect of titanium oxide coating on the performance characteristics of the bio-diesel-fueled engine. The layer of thermal coating was characterized by alumina-titania (Al_2O_3/TiO_2) plasma coated on to the base of NiCrAl. The experiments were conducted with a single cylinder, four stroke, and direct injected diesel engine and the results showed an increase in brake thermal efficiency and a decrease in brake specific fuel consumption for titanium oxide coated piston [58].

Ravikumar and Senthilkumar investigated radish (Raphanus sativus) oil Methyl Ester in a single cylinder DI diesel engine with and without coating. The effect of biodiesel in a thermal barrier coating engine was studied comparing diesel and radish oil methyl ester used as fuels. In this study the piston crown surface, valves and cylinder head of a diesel engine were coated with a ceramic material-TiO_2 (Figure 20). The application of TiO_2 coating slightly increased Specific Fuel Consumption (SFC), emissions of CO, smoke density, HC and NOx and decreased brake thermal efficiency. From the above experimental results, it was demonstrated that B25 with TiO_2-coated mode of engine operation gave better performance and lower emission characteristics including NOx, without requiring any major modification of engine [59].

Figure 20. Photographic view of 450 lm TiO_2 coated piston, valves, and cylinder head (Reproduced from [59], with the permission of AIP Publishing).

5. Conclusions

This work has been useful in assessing the possible catalytic pathways in the production of biodiesel, exploring particularly the use of titanium dioxide as catalyst. By evaluating the different parameters, among them type and percentage of titania-based catalyst, temperature, time and alcohol/oil ratio, it was possible to evaluate the optimized conditions leading to the best conversion yields, in batch conditions.

Furthermore, this work focused on the study of the different strategies to conduct the transesterification reaction mediated by titanium dioxide as catalyst within microreactors.

Optimized conditions in continuous flow, resulted improved by modifying parameters such as temperature, pressure, and residence time and also allowed a possible recovery and reuse of glycerol.

The use of micro-technologies in biodiesel production and the development of micro-reactors, capable of making possible unpracticable chemical transformations using traditional techniques, allows a reduction in production costs and a greater protection of the environment.

Author Contributions: Writing—Review and Editing, C.C, L.D., R.L.

Funding: This research received no external funding.

Acknowledgments: This work was supported by the Intervento cofinanziato dal Fondo di Sviluppo e Coesione 2007-2013—APQ Ricerca Regione Puglia "Programma regionale a sostegno della specializzazione intelligente e della sostenibilità sociale ed ambientale—FutureInResearch".

Conflicts of Interest: The authors declare no conflict of interest.

References

1. Thanh, L.T.; Okitsu, K.; Van Boi, L. Maeda, Y. Catalytic Technologies for Biodiesel Fuel Production and Utilization of Glycerol: A Review. *Catalysts* **2012**, *2*, 191–222. [CrossRef]
2. Glisic, S.B.; Pajnik, J.M.; Orlović, A.M. Process and techno-economic analysis of green diesel production from waste vegetable oil and the comparison with ester type biodiesel production. *Appl. Energy* **2016**, *170*, 176–185. [CrossRef]
3. Bobadilla Corral, M.; Lostado Lorza, R.; Escribano García, R.; Somovilla Gómez, F.; Vergara González, E.P. An Improvement in Biodiesel Production from Waste Cooking Oil by Applying Thought Multi-Response Surface Methodology Using Desirability Functions. *Energies* **2017**, *10*, 130. [CrossRef]
4. Bavykin, D.V.; Walsh, F.C. Titanate and Titania Nanotubes: Synthesis. *RSC Nanosci. Nanotechnol.* **2009**. [CrossRef]
5. Bartl, M.H.; Boettcher, S.W.; Frindell, K.L.; Stucky, G.D. Molecular assembly of function in titania-based composite material system. *Acc. Chem. Res.* **2005**, *38*, 263–271. [CrossRef] [PubMed]
6. Linsebigler, A.L.; Lu, G.; Yates, J.T. Photocatalysis on TiO_2 Surfaces: Principles, Mechanisms, and Selected Results. *Chem. Rev.* **1995**, *95*, 735–758. [CrossRef]
7. Siling, M.I.; Laricheva, T.N. Titanium compounds as catalysts for esterification and transesterification. *Russ. Chem. Rev.* **1996**, *65*, 279–286. [CrossRef]
8. Carlucci, C.; Xu, H.; Scremin, B.F.; Giannini, C.; Sibillano, T.; Carlino, E.; Videtta, V.; Gigli, G.; Ciccarella, G. Controllable One-Pot Synthesis of Anatase TiO_2 Nanorods with the Microwave-Solvothermal Method. *Sci. Adv. Mater.* **2014**, *6*, 1668–1675. [CrossRef]
9. Bagheri, S.; Julkapli, N.M.; Hamid, S.B.A. Titanium Dioxide as a Catalyst Support in Heterogeneous Catalysis. *Sci. World J.* **2014**, *2014*, 727496. [CrossRef]
10. Noda, L.K.; de Almeida, R.M.; Probst, L.F.D.; Gonçalves, N.S. Characterization of sulfated TiO_2 prepared by the sol–gel method and its catalytic activity in the *n*-hexane isomerization reaction. *J. Mol. Catal. A Chem.* **2005**, *225*, 39–46. [CrossRef]
11. Refaat, A. Biodiesel production using solid metal oxide catalysts. *Int. J. Environ. Sci. Technol.* **2011**, *8*, 203–221. [CrossRef]
12. Noda, L.K.; de Almeida, R.M.; Gonçalves, N.S.; Probst, L.F.D.; Sala, O. TiO_2 with a high sulfate content -thermogravimetric analysis, determination of acid sites by infrared spectroscopy and catalytic activity. *Catal. Today* **2003**, *85*, 69–74. [CrossRef]
13. Cui, H.; Dwight, K.; Soled, S.; Wold, A. Surface Acidity and Photocatalytic Activity of Nb_2O_5/TiO_2 Photocatalysts. *J. Solid State Chem.* **1995**, *115*, 187–191. [CrossRef]
14. Li, G. Surface modification and charachterizations of TiO_2 nanoparticle. *Surf. Rev. Lett.* **2009**, *16*, 149–151. [CrossRef]
15. Gardy, J.; Hassanpour, A.; Lai, X.; Ahmed, M.H. Synthesis of $Ti(SO_4)O$ solid acid nano-catalyst and its application for biodiesel production from used cooking oil. *Appl. Catal. A Gen.* **2016**, *527*, 81–95. [CrossRef]

16. Gardy, J.; Hassanpour, A.; Laia, X.; Ahmed, M.H.; Rehan, M. Biodiesel production from used cooking oil using a novel surface functionalised TiO$_2$ nano-catalyst. *Appl. Catal. B Environ.* **2017**, *207*, 297–310. [CrossRef]
17. Zhao, H.; Jiang, P.; Dong, Y.; Huang, M.; Liu, B. A high-surface-area mesoporous sulfated nano-titania solid superacid catalyst with exposed (101) facets for esterification: Facile preparation and catalytic performance. *New J. Chem.* **2014**, *38*, 4541–4548. [CrossRef]
18. Ropero-Vega, J.L.; Aldana-Péreza, A.; Gómez, R.; Nino-Gómez, M.E. Sulfated titania [TiO$_2$/SO$_4{}^{2-}$]: A very active solid acid catalyst for the esterification of free fatty acids with ethanol. *Appl. Catal. A Gen.* **2010**, *379*, 24–29. [CrossRef]
19. Li, X.; Huang, W. Synthesis of Biodiesel from Rap Oil over Sulfated Titania-based Solid Superacid Catalysts. *Energy Sources Part A* **2009**, *31*, 1666–1672. [CrossRef]
20. de Almeida, R.M.; Noda, L.K.; Gonçalves, N.S.; Meneghetti, S.M.P.; Meneghetti, M.R. Transesterification reaction of vegetable oils, using superacid solfate TiO$_2$-base catalysts. *Appl. Catal. A Gen.* **2008**, *347*, 100–105. [CrossRef]
21. Chen, H.; Peng, B.; Wang, D.; Wang, J. Biodiesel production by the transesterification of cottonseed oil by solid acid catalysts. *Front. Chem. Eng. China* **2007**, *1*, 11–15. [CrossRef]
22. Oprescu, E.E.; Velea, S.; Doncea, S.; Radu, A.; Stepan, E.; Bolocan, I. Biodiesel from Algae Oil with High Free Fatty Acid over Amphiphilic Solid Acid Catalyst. *Chem. Eng. Trans.* **2015**, *43*, 595–600. [CrossRef]
23. Boffito, D.C.; Crocellà, V.; Pirola, C.; Neppolian, B.; Cerrato, G.; Ashokkumar, M.; Bianchi, C.L. Ultrasonic enhancement of the acidity, surface area and free fatty acids esterification catalytic activity of sulfated ZrO$_2$-TiO$_2$ systems. *J. Catal.* **2013**, *297*, 17–26. [CrossRef]
24. Peng, B.-X.; Shu, Q.; Wang, J.-F.; Wang, G.-R.; Wang, D.-Z.; Han, M.-H. Biodiesel production from waste oil feedstocks by solid acid catalysis. *Process. Saf. Environ. Prot.* **2008**, *86*, 441–447. [CrossRef]
25. Shao, G.N.; Sheikh, R.; Hilonga, A.; Lee, J.E.; Park, Y.-H.; Kim, H.T. Biodiesel production by sulfated mesoporous titania-silica catalysts synthesized by the sol–gel process from less expensive precursors. *Chem. Eng. J.* **2013**, *215–216*, 600–607. [CrossRef]
26. Embong, N.H.; Maniam, G.P.; Rahim, M.H.A. Biodiesel Preparation from Decanter Cake with Solid Acid Catalyst. *Int. J. Chem. Environ. Eng.* **2014**, *5*, 294–296.
27. Wang, K.; Jiang, J.; Si, Z.; Liang, X. Biodiesel production from waste cooking oil catalyzed by solid acid SO$_4{}^{2-}$/TiO$_2$/La^{3+}. *J. Renew. Sustain. Energy* **2013**, *5*, 052001. [CrossRef]
28. Li, Y.; Zhang, X.-D.; Sun, L.; Zhang, J.; Xu, H.-P. Fatty acid methyl ester synthesis catalyzed by solid superacid catalyst SO$_4{}^{2-}$/ZrO$_2$-TiO$_2$/La^{3+}. *Appl. Energy* **2010**, *87*, 156–159. [CrossRef]
29. Anuradha, S.; Raj, K.J.A.; Vijayaraghavan, V.R.; Viswanathan, B. Sulfated Fe$_2$O$_3$-TiO$_2$ catalysed transesterification of soybean oil to biodiesel. *Indian J. Chem.* **2014**, *53*, 1493–1499.
30. Emeji, I.C.; Afolabi, A.S.; Abdulkareem, A.S.; Kalala, J. Characterization and Kinetics of Biofuel Produced from Waste Cooking Oil. In Proceedings of the World Congress on Engineering and Computer Science, San Francisco, CA, USA, 21–23 October 2015; Volume II.
31. Madhuvilakku, R.; Piraman, S. Biodiesel synthesis by TiO$_2$-ZnO mixed oxide nanocatalyst catalyzed palm oil transesterification process. *Bioresour. Technol.* **2013**, *150*, 55–59. [CrossRef]
32. Sithole, T.; Meijboom, R.; Jalama, K. Biodiesel Production from Waste Vegetable Oil over MgO/TiO$_2$ catalyst. *IJESIT* **2013**, *2*, 189–194.
33. Mguni, L.L.; Meijboom, R.; Jalama, K. Biodiesel Production over nano-MgO Supported on Titania. *Int. J. Chem. Mol. Nuclear Mater. Metall. Eng.* **2012**, *6*, 380–384.
34. Wen, Z.; Yu, X.; Tu, S.-T.; Yan, J.; Dahlquist, E. Biodiesel production from waste cooking oil catalyzed by TiO$_2$-MgO mixed oxides. *Bioresour. Technol.* **2010**, *101*, 9570–9576. [CrossRef] [PubMed]
35. Kawashima, A.; Matsubara, K.; Honda, K. Development of heterogeneous base catalysts for biodiesel production. *Bioresour. Technol.* **2008**, *99*, 3439–3443. [CrossRef] [PubMed]
36. Salinas, D.; Guerrero, S.; Araya, P. Transesterification of canola oil on potassium-supported TiO$_2$ catalysts. *Catal. Commun.* **2010**, *11*, 773–777. [CrossRef]
37. Salinas, D.; Araya, P.; Guerrero, S. Study of potassium-supported TiO$_2$ catalysts for the production of biodiesel. *Appl. Catal. B Environ.* **2012**, *117–118*, 260–267. [CrossRef]
38. Martínez-Klimova, E.; Hernández-Hipólito, P.; Klimova, T.E. Biodiesel Production with Nanotubular Sodium Titanate Doped with Potassium as a Catalyst. *MRS Adv.* **2016**, *1*, 415–420. [CrossRef]

39. Colella, M.; Carlucci, C.; Luisi, R. Supported Catalysts for Continuous Flow Synthesis. *Top. Curr. Chem.* **2018**, *376*, 46. [CrossRef]
40. Fanelli, F.; Parisi, G.; Degennaro, L.; Luisi, R. Contribution of microreactor technology and flow chemistry to the development of green and sustainable synthesis. *Beilstein J. Org. Chem.* **2017**, *13*, 520–542. [CrossRef]
41. Degennaro, L.; Carlucci, C.; De Angelis, S.; Luisi, R. Flow Technology for Organometallic-Mediated Synthesis. *J. Flow Chem.* **2016**, *6*, 136–166. [CrossRef]
42. Degennaro, L.; Fanelli, F.; Giovine, A.; Luisi, R. External trapping of halomethyllithium enabled by flow microreactors. *Adv. Synth. Catal.* **2014**, *357*, 21–27. [CrossRef]
43. Tran, D.-T.; Chang, J.-S.; Lee, D.-J. Recent insights into continuous-flow biodiesel production via catalytic and non-catalytic transesterification processes. *Appl. Energy* **2017**, *185*, 376–409. [CrossRef]
44. Tiwari, A.; Rajesh, V.M.; Yadav, S. Biodiesel production in micro-reactors: A review. *Energy Sustain. Dev.* **2018**, *43*, 143–161. [CrossRef]
45. Khana, Y.; Marina, M.; Viinikainena, T.; Lehtonena, J.; Puurunena, R.L.; Karinena, R. Structured microreactor with gold and palladium on titania: Active, regenerable and durable catalyst coatings for the gas-phase partial oxidation of 1-butanol. *Appl. Catal. A Gen.* **2018**, *562*, 173–183. [CrossRef]
46. Barot, S.; Bandyopadhyay, R.; Joshi, S.S. Catalytic Conversion of Jatropha Oil to Biofuel Over Titania, Zirconia, and Ceria Loaded Amorphous Alumino-Silicate Catalysts. *Environ. Prog. Sustain. Energy* **2017**, *36*, 749–757. [CrossRef]
47. McNeff, C.V.; McNeff, L.C.; Yan, B.; Nowlan, D.T.; Rasmussen, M.; Gyberg, A.E.; Krohn, B.J.; Fedie, R.L.; Hoye, T.R. A continuous catalytic system for biodiesel production. *Appl. Catal. A Gen.* **2008**, *343*, 39–48. [CrossRef]
48. Krohn, B.J.; McNeff, C.V.; Yan, B.; Nowlan, D. Production of algae-based biodiesel using the continuous catalytic Mcgyan® process. *Bioresour. Technol.* **2011**, *102*, 94–100. [CrossRef]
49. Baroutian, S.; Aroua, M.K.; Raman, A.A.A.; Sulaiman, N.M.N. A packed bed membrane reactor for production of biodiesel using activated carbon supported catalyst. *Bioresour. Technol.* **2011**, *102*, 1095–1102. [CrossRef]
50. Len, C.; Delbecq, F.; Cara Corpas, C.; Ruiz Ramos, E. Continuous Flow Conversion of Glycerol into Chemicals: An Overview. *Synthesis* **2018**, *50*, 723–741. [CrossRef]
51. Bhanuchander, P.; Priya, S.S.; Kumar, V.P.; Hussain, S.; Pethane Rajan, N.; Bhargava, S.K.; Chary, K.V.R. Direct Hydrogenolysis of Glycerol to Biopropanols over Metal Phosphate Supported Platinum Catalysts. *Catal. Lett.* **2017**, *147*, 845–855. [CrossRef]
52. Liebig, C.; Paul, S.; Katryniok, B.; Guillon, C.; Couturier, J.L.; Dubois, J.L.; Dumeignil, F.; Hoelderich, W.F. Glycerol conversion to acrylonitrile by consecutive dehydration over WO_3/TiO_2 and ammoxidation over Sb-(Fe,V)-O. *Appl. Catal. B Environ.* **2013**, *132–133*, 170–182. [CrossRef]
53. Jeryrajkumar, L.; Anbarasu, G.; Elangovan, T. Effects on Nano Additives on Performance and Emission Characteristics of Calophyllim inophyllum Biodiesel. *Int. J. ChemTech Res.* **2016**, *9*, 210–219.
54. Venu, H.; Madhavan, V. Effect of nano additives (titanium and zirconium oxides) and diethyl ether on biodiesel-ethanol fuelled CI engine. *J. Mech. Sci. Technol.* **2016**, *30*, 2361–2368. [CrossRef]
55. D'Silva, R.; Vinoothan, K.; Binu, K.G.; Bhat, T.; Raju, K. Effect of Titanium Dioxide and Calcium Carbonate Nanoadditives on the Performance and Emission Characteristics of C.I. Engine. K. *J. Mech. Sci. Autom.* **2016**, *6*, 28–31. [CrossRef]
56. Prabhu, L.; Satish Kumar, S.; Andrerson, A.; Rajan, K. Investigation on performance and emission analysis of TiO_2 nanoparticle as an additive for biodiesel blends. *J. Chem. Pharm. Sci.* **2015**, *7*, 408–412.
57. Fangsuwannarak, K.; Triratanasirichai, K. Improvements of Palm Biodiesel Properties by Using Nano-TiO_2 Additive, Exhaust emission and Engine Performance. *Rom. Rev. Precis. Mech. Opt. Mechatron.* **2013**, *43*, 111–118.

58. Naveen, P.; Rajashekhar, C.R.; Umashankar, C.; Rajashekhar Kiragi, V. Effect of Titanium Oxide Coating on Performance Characteristics of Bio-Diesel (Honge) Fuelled C. I. Engine. *IJMER* **2012**, *2*, 2825–2828.
59. Ravikumar, V.; Senthilkumar, D. Reduction of NOx emission on NiCrAl-Titanium Oxide coated direct injection diesel engine fuelled with radish (Raphanus sativus) biodiesel. *J. Renew. Sustain. Energy* **2013**, *5*, 063121. [CrossRef]

© 2019 by the authors. Licensee MDPI, Basel, Switzerland. This article is an open access article distributed under the terms and conditions of the Creative Commons Attribution (CC BY) license (http://creativecommons.org/licenses/by/4.0/).

MDPI
St. Alban-Anlage 66
4052 Basel
Switzerland
Tel. +41 61 683 77 34
Fax +41 61 302 89 18
www.mdpi.com

Catalysts Editorial Office
E-mail: catalysts@mdpi.com
www.mdpi.com/journal/catalysts

www.ingramcontent.com/pod-product-compliance
Lightning Source LLC
LaVergne TN
LVHW071954080526
838202LV00064B/6748